Contested Markets, Contested Cities

Markets are at the origin of urban life as places for social, cultural and economic encounter evolving over centuries. Today, they have a particular value as mostly independent, non-corporate and often informal work spaces serving millions of the most vulnerable communities across the world. At the same time, markets have become fashionable destinations for 'foodies' and middle class consumers and tourists looking for authenticity and heritage. The confluence of these potentially contradictory actors and their interests turns markets into "contested spaces".

Contested Markets, Contested Cities provides an analytical and multi-disciplinary framework within which specific markets from Mexico City, Buenos Aires, Santiago de Chile, Quito, Sofia, Madrid, London and Leeds (UK) are explored. This pioneering and highly original work examines public markets from a perspective of contestation looking at their role in processes of gentrification but also in political mobilisation and urban justice.

Sara González is Associate Professor in the School of Geography at the University of Leeds, UK where she teaches on critical urban geography courses and leads the research group on Social Justice, Cities and Citizenship. She has published work in international journals on the political and economic transformation in cities, the neoliberalisation of urban policies, gentrification and grassroots contestation of these processes. Between 2012 and 2016 she was the Principal Investigator in Leeds of the EU-funded Contested Cities network, bringing together more than 40 researchers across Europe and Latin America. Between 2006 and 2016 she was part of the editorial collective of the open-access *ACME: An International Journal for Critical Geographies*. She favours participatory action research methodologies and has been a very active member of a campaign to support her local market in Leeds.

T0175076

Routledge Studies in Urbanism and the City

This series offers a forum for original and innovative research that engages with key debates and concepts in the field. Titles within the series range from empirical investigations to theoretical engagements, offering international perspectives and multidisciplinary dialogues across the social sciences and humanities, from urban studies, planning, geography, geohumanities, sociology, politics, the arts, cultural studies, philosophy and literature.

For a full list of titles in this series, please visit www.routledge.com/series/RSUC

Contested Markets, Contested Cities

Gentrification and Urban Justice in Retail Spaces

Edited by
Sara González

Routledge
Taylor & Francis Group

LONDON AND NEW YORK

First published 2018 by Routledge

2 Park Square, Milton Park, Abingdon, Oxfordshire OX14 4RN
52 Vanderbilt Avenue, New York, NY 10017

Routledge is an imprint of the Taylor & Francis Group, an informa business

First issued in paperback 2019

British Library Cataloguing in Publication Data
A catalogue record for this book is available from the British Library

Library of Congress Cataloging in Publication Data
A catalog record for this book has been requested

ISBN: 978-1-138-21748-5 (hbk)
ISBN: 978-0-367-87844-3 (pbk)

Typeset in Times New Roman
by Taylor & Francis Books

Contents

Illustrations

Figures

Tables

Contributors

María Jesús Arce Sánchez is an architect with a Master's degree in Urban Projects from the School of Architecture of the Pontificia Universidad Católica de Chile. Her Master's thesis analyses projects of urban renewal related to La Vega Central market and its neighbourhood, focusing on their inclusive socio-cultural characteristics. The thesis was carried out as part of the Chilean National Research Council project (Fondecyt, no. 1120823), entitled: 'Lo Público y lo Privado en la Producción de Espacios Públicos Vitales (The public and the private in the production of vital public spaces), led by Elke Schlack, for which she carried out interviews, socio-spatial mapping and participant observation. She is currently working in the private sector developing urban research and engaging with city planning.

María Ximena Arqueros is a Lecturer in the Department of Economy, Development and Agrarian Planning, Faculty of Agronomy at the University of Buenos Aires, Argentina, where she teaches issues related to environmental conflicts, rural sociology, community engagement and agroecology. She has integrated several engagement and research projects analysing issues related to urban agroecology, urban agriculture, food sovereignty, community intervention and university engagement partnerships. Since 1997 she has been a coordinator of the Programme of University engagement in Community and School Gardening (PEUHEC). Her research studies socio-spatial transformations in the context of the global expansion of capitalism, analysing the way in which organisational social practices and public policies affect territorial reconfigurations.

Erika Bedón is a PhD candidate in Anthropology from the University of Rovira i Virgili, Spain. She holds a Master's in Social Sciences and a Bachelor's degree in Anthropology from FLACSO-Ecuador. She is an Associate Researcher in the Department of Anthropology, History and Humanities of FLACSO-Ecuador, where she is member of the Memory Studies research group. More widely she is also member of the Working Group on Indigenous People and Urban Space of CLACSO, the Latin American Council of Social Sciences. She has worked on issues of urban memory and indigenous migration and published specifically about San Roque market on subjects such as collective memory, heritage and migrant children strategies of survival in and around the markets

Luz de Lourdes Cordero Gómez del Campo holds a degree in Business Administration, specialising in market systems, and a degree in Geography. From 2013–2016 she was part of the research group Contested Cities at the National Autonomous University of Mexico, a network funded by the European Union with universities from Europe and Latin America. She is completing a Master's degree in Geography at the Universidad Nacional Autónoma de México with a scholarship from the Mexican National Council of Science and Technology. Her research interests are commercial gentrification, specifically in regard to gourmet markets, staged urban spaces and city branding.

Gloria Dawson is a writer and an independent researcher. She holds an MA in Social and Cultural Geography from the University of Leeds and her interests include precarious housing, the relationship between housing and work and gentrification (especially in terms of retail, food, art and culture).

Victor Delgadillo is Professor in the Department of Urban Affairs and Political Science of the Autonomous University of Mexico City and in the PhD programme on Urbanism of the National Autonomous University of Mexico. There he teaches courses about city transformations in the neoliberal era, urban planning, urban heritage and housing in urban central areas of Latin American cities. His research mainly deals with gentrification processes in Mexico City and the commodification of urban heritage in inner city areas in Latin America. From 2012 to 2016 he was the coordinator of one of the two research groups in Mexico City within the Contested Cities network.

Stoyanka Andreeva Eneva is a PhD research fellow in the Department of Political Science & International Relations of the Autonomous University of Madrid. Her research focuses on commercial gentrification processes in public markets in both Madrid and Sofia. She has published several co-authored articles which compare market transformation models in Madrid with other cities such as Barcelona and Brasilia. Her research focuses on the role of ethnic diversity in marketplaces and on the mobility of urban policies.

Nela Lena Gallardo Araya is a Lecturer in the Department of Economy, Development and Agrarian Planning, Faculty of Agronomy at the University of Buenos Aires, Argentina, where she teaches on sociology, rural engagement and agroecology. She is also a member of the group of Anthropology, City and Nature in the Institute of Gino Germani at the Faculty of Social Sciences and has published on urban agriculture, agroecology, food sovereignty and rural engagement. Since 2001 she has been a

Coordinator of the Program of University Engagement in Community and School Gardening (PEUHEC). Since 2016 she has been part of the editorial collective of the journal *Practicas de Oficio*. She supports urban agroecology and has been a very active member of agriculture networks in Buenos Aires.

Eva García Pérez is an urbanist-architect who graduated from Escuela Técnica Superior de Arquitectura de Madrid (ETSAM), and continued her training in participatory planning in Facultad Latinoamericana de Ciencias Sociales (FLACSO-Argentina). She has been a member of the the Observatorio Metropolitano de Madrid (OM) as a scholar activist and has also been part of the Contested Cities project (2012–2016), where she developed her research around gentrification processes, real estate market dynamics and urban segregation and inequality. Since September 2015 she has been working as an advisor in the Madrid Municipal government without leaving the grassroots anti-gentrification activism. She is also writing a PhD thesis looking at urban transformations in Madrid during the crisis following the Spanish 'housing bubble'.

Norma Angélica Gómez Méndez is full-time Professor in the Political Science and Urban Administration Academy at the Autonomous University of Mexico City, where she teaches methodology in social sciences. She has published articles analysing the informal economy, social and political relations and organisations in street commerce in Mexico City, the different types of street vendors and the problem of gender in this kind of commerce. She joined the Contested Cities project in 2014.

Sara González is Associate Professor in the School of Geography at the University of Leeds, where she teaches on critical urban geography courses and leads the research group on Social Justice, Cities and Citizenship. She has published work in international journals on the political and economic transformation of cities, the neoliberalisation of urban policies, gentrification and grassroots contestation of these processes. Between 2012 and 2016 she was the Principal Investigator in Leeds of the EU-funded Contested Cities network, bringing together more than 40 researchers across Europe and Latin America. Between 2006 and 2016 she was part of the editorial collective of the open-access *ACME: An International Journal for Critical Geographies*. She favours participatory action research methodologies and has been an active member of a campaign to support her local market in Leeds.

Victoria Habermehl is a Research Associate in the Urban Institute at the University of Sheffield, working on the ESRC 'Whose Knowledge Matters?' project and is part of the Mistra Urban Futures network. She has a PhD in Geography from the University of Leeds, focused on organising in-against-and-beyond crisis in Buenos Aires, Argentina, through the economy, state and territory. Between 2012 and 2016 she was part of the Contested Cities network. Victoria's research focuses on everyday life in cities and the antagonistic processes involved in urban politics, embodied in social reproduction, neighbourhood organising, gentrification and displacement. Her research to date has focused on Latin American cities, and in particular in Buenos Aires, Argentina where she researched how the economy is understood and reshaped through crisis, narratives of economic informality and everyday economic practices such as economic solidarity initiatives, autogestion and popular economy.

Eduardo Kingman Garcés holds a PhD in Urban Anthropology from the Universitat Rovira i Virgili, Spain. He is Research Professor of Anthropology at FLACSO-Ecuador where he is part of the collective of studies of social memory. He is a specialist in urban anthropology and work of memory of subaltern groups based on life histories. He is the author of several books such *as La ciudad y los otros: Quito 1860–1940. Higienismo, ornato y policia (The city and the others: Quito 1860–1940, Hygienist Policies, Ornament and Police)*. He has also published widely on popular culture, memory, security, heritage and specifically about how these issues confluence in the San Roque market and neighbourhood.

Vincenzo Maiello is an architect who has worked in Italy, Portugal and Spain. He holds two Master's degrees, one in Urban Projects from the Faculty of Architecture of ISCTE (Lisbon) and the other in Development Cooperation in Precarious Human Settlements from the Polytechnic University of Madrid. Since 2010 he has been working in the field of International Development cooperation. He is currently Project Manager at the Madrid-based Office of the NGO Architects Without Borders Spain (Arquitectura Sin Fronteras España), where he coordinates projects in several countries of Africa, Latin America and in the Caribbean as well as local actions in Madrid to promote the 'Right to the City', for instance in defense of the city's municipal markets.

Elvira Mateos Carmona is a researcher at Autonomous University of Madrid. Her main interests are the construction of subjectivities in articulation with processes of neoliberalisation in urban centres, as well as the discursive analysis of conflicts related to spatial injustice. Between 2012 and 2016 she has been a member of the Contested Cities project. She is also part of Urban Studies and Social Theory Research group at the Autonomous University of Madrid. She is presently working on her PhD thesis, focusing on the transformation of municipal markets in the centre of Madrid, under the supervision of Professor Michael Janoschka.

Penny Rivlin is a Lecturer in the School of Media and Communication at the University of Leeds, where she teaches courses on media, society and culture and communication research methods. With disciplinary roots in sociology and cultural studies, Penny's doctoral research explored state-sponsored food waste and eco-campaigns in relation to the gendered, classed dimensions of household sustainable consumption and practice. She has worked as a researcher on a range of UK-based funded projects that explore the relationships between communities, place, the digital and sustainable cultures, most recently working with Sara González on the British Academy/Sadler-funded *Leeds Voices: Communicating Superdiversity in the Market.*

Alejandro Rodríguez Sebastián is an architect and holds a Master's degree in Urban and Land Planning from the Polytechnic University of Madrid, an institution where he has worked as a researcher for different projects related to the fields of urban regeneration and urban vulnerability. He is coordinator of the 'CF+S Library' (an organisation linked to UN-HABITAT in Spain) and co-editor of the research journal *Territorios en Formación* (*Territories in Formation*). Alejandro is currently working on his PhD at the Polytechnic University of Madrid, studying the transformation of Madrid's traditional markets and its relationship with the city's socio-spatial evolution.

Luis Alberto Salinas Arreortua is a researcher at the Geography Institute of the National Autonomous University of Mexico, where he teaches Political Geography at undergraduate level. He also teaches a course on the Neoliberal City on the PhD in Urban Studies. He has published several articles in international journals as well as book chapters. Between 2012 and 2016 he took part in the Contested Cities network. His research topics are housing policies, gentrification and the neoliberal city. He received the recognition for the best doctoral thesis in Social Geography in 2013 at his university and since 2015 he has been a member of the Mexican National Research Council, CONACYT, México.

Elke Schlack is Assistant Professor in the School of Architecture at the Pontificia Universidad Católica de Chile and in the 'Campus Creativo' (Creative Campus) at the Andres Bello University in Chile. She has led the Urban Projects Masters Programme and is a researcher at the Laboratorio de Movilidad (Mobility Laboratory) at the Pontificia Universidad Católica de Chile. She has published in international journals and book chapters on urban theory, design morphology, public policy and urban regulations related to urban renewal and public space. Her research has been funded by the Chilean Commission for Scientific and Technological Research (CONICYT) and concerns the realm of urban space, focusing on the public sphere, commerce and socially inclusive renewal processes. She is author and editor of the book *POPS – el uso público en el espacio urbano* (public use in urban space), a critical approach to incentive zoning in Chile. She works internationally in collaboration with Neil Turnbull on urban research projects.

Neil Turnbull is an architect and academic at the Faculty of Architecture and Urbanism of the University of Chile from 2012–16. He has participated in state-funded research projects and published work on the processes and social impact of urban regeneration. He is currently engaged in a PhD studentship, funded by the Economic and Social Research Council (UK) at the School of Geography and Planning of Cardiff University. This work involves research into the transformative opportunities of local governance with a focus on micro geographies of communal and public space. He works internationally in collaboration with Elke Schlack on urban research projects.

Preface

This book has been a true collective effort bringing together the individual passion that each of the contributors has for markets. Although this is an edited collection with many different authors, the research that underpins it has been developed collaboratively through a multitude of emails, written comments to drafts and many Skype and face to face meetings over four years.

More specifically, the book is the result of the CONTESTED CITIES project, an EU funded Marie Curie International research staff exchange scheme (Reference FP7-PEOPLE-PIRSES-GA-2012–318944) between 2012 and 2016. This project, titled 'Contested spatialities of urban neoliberalism – Dialogues between emerging spaces of citizenship in Europe and Latin America' aimed at exchanging research on cities between Latin American and European colleagues. The project was led by Dr Michael Janoschka who at the time was working at the Autonomous University of Madrid, and involved research groups from Madrid itself, Buenos Aires, Santiago de Chile, Rio de Janeiro, Mexico City and Leeds, where I was the local coordinator. When we started, the project was focused on analysing gentrification, the neoliberalisation of cities and practices of resistance to these processes from the grassroots. I was by then interested in the gentrification of traditional public markets through my experience first as a customer in Leeds Kirkgate Market and then as a campaigner with the group Friends of Leeds Kirkgate Market which I co-founded in April 2010. This subsequently became an academic interest and with my colleague in Leeds Paul Waley we wrote an article on the gentrification of markets.

The Contested Cities network functioned through distant email communications but more importantly through periodic face to face meetings and reciprocal exchange visits between the European and Latin American universities. By our second meeting in Santiago de Chile in April 2014 some of us had started to share our interest in researching markets. Eva García, Nela Gallardo, Vicky Habermehl, Victor Delgadillo, Lil (Luz Cordero) and I (all authors in this book) were in particular involved and we decided to start a 'markets sub-group' within the network. During our meeting in Santiago de Chile we visited La Vega market (featured in the book) led by gentrification expert and project colleague Ernesto López. A few days later I met Neil Turnbull and Elke Schlack who were both working on retail gentrification in Santiago de Chile and interested in processes of transformation in La Vega, and they joined our markets subgroup.

At our next meeting in Leeds in September 2014 we presented preliminary research on case studies from our cities and also enrolled Luis Salinas working on markets in Madrid but originally from Mexico City. Ana Cabrera, studying for a PhD in Leeds on Maya settlements in Yucatán and Uriel Martínez, working on Mexico City also joined us but sadly they had to later concentrate on their PhD theses and their work did not make it into the book. It was in Leeds that we decided to write a book together and at our next meeting in Mexico City in April 2015 we presented draft chapters commenting and reviewing each other's work. By then we had also enrolled Elvira Mateos, from the Contested Cities Madrid team, who was working on processes of transformation in the public markets of the city from an anthropological perspective. While in Mexico City, we all enjoyed a visit to the La Merced markets (featured in the book) led by Victor Delgadillo where we met and talked to traders and decision makers. Interestingly we also had a group interview with a high level official in Mexico City with responsibilities over markets.

In our final network meeting in Rio de Janeiro in December 2015 we reviewed again our draft chapters and discussed the main connecting themes of the book to be developed in the introduction. By then we had already agreed a contract for a book with Routledge. In the book review process, the reviewers suggested that we should expand our case studies to other Latin American and Easter European Cities. Colleagues in Quito had recently joined the Contested Cities network and so through them we enrolled Erika Bedon and Eduardo Kingman who had already years of research expertise in the San Roque market there. A recently joined PhD student in the Contested Cities Madrid team, Tania (Stoyanka) Eneva had being doing research on the Women's Market in Sofia so we invited her along too.

In July 2016 Contested Cities had its final meeting but this time it was an open conference where we invited people from all over the world to present their work on the areas that we had been developing for years within our network. We had huge success attracting papers on markets and retail gentrification and we had several sessions dedicated to discuss these issues. I finally had the pleasure to meet in person some authors of the book such as Tania Eneva and Alejandro Rodríguez.

In parallel to this book, there are also various projects that have influenced our work and deserve special mention. Victor Delgadillo, contributor to this book, has been coordinating several publications in Spanish on the same topic that have appeared in the Mexican academic journals *Alteridades* and *Ciudades* and where some of the authors of this book have also published in Spanish. This is very important work as it expands the readership of our research. In 2014, I was awarded a scholar-activist award by the Antipode

Foundation to work with market campaigners across the UK and conduct further research on gentrification. Thanks to this funding I employed Gloria Dawson, co-author of the London chapter in this book and who helped enormously in the last phases of this book as copy editor, proof-reader and indexer. In 2015 I got involved with other colleagues at the University of Leeds in an interdisciplinary project titled *Leeds Voices* where we looked at Leeds Kirkgate Market as a site to understand superdiversity in cities. In this project, Penny Rivlin, co-author in the Leeds chapter, conducted interviews with market traders and made multiple observations.

I am hugely indebted to the excellent work of all the contributors of this book. As mentioned earlier, this is a collective book which builds on the individual passion and expertise of a team of authors. The author biographies reveal very diverse skills sets and disciplinary homes that give this book enormous richness. Communication within the Contested Cities network was always mainly in Spanish and we always enjoyed noticing the differences of our common language across various countries.

There are several people that deserve a special acknowledgement for their role in making this book possible. Neil Turnbull and Juan Pablo Henríquez translated six of the chapters from Spanish, making it possible for their authors to be read in English. It is worth mentioning that creating this book has been a bilingual journey. Special acknowledgements also go to Carl Waterston who was in charge of the finances of Contested Cities in Leeds (which paid for meetings, flights, translations and copy-editing), Jan O'Brien the wife of the late photographer Colin O'Brien for allowing us to use one of his photos in Chapter 4 and the book cover free of charge and again Gloria Dawson for the detailed work and patience in putting the book manuscript together. I am also grateful to Michael Janoschka, principal investigator of Contested Cities, who trusted me to get involved in this project and always supported the line of research on markets. Final thanks go to my partner Stuart Hodkinson and our children Oihana and Diego for making it possible for me to travel around in Latin America and putting up with my market obsession.

We dedicate this book to the traders and market campaigners that shared with us their experiences and who struggle every day for their livelihoods and to keep markets as open and affordable public spaces.

Sara Gonzalez, Leeds, April 2017

1 Introduction

Studying markets as spaces of contestation

Sara González

Traditional markets, where food and other goods are sold on the streets, in covered regulated spaces or in informal settings, are still serving millions of people across the world despite the advance of corporate and globalised supermarkets. They are not only important spaces for exchange in the local economy but also for social interaction, and in particular they are essential to the most vulnerable communities in our cities, from migrant workers, ethnic minorities, the elderly and the poor. At the same time, in recent decades many markets across the world have been rediscovered as tourist attractions, food meccas and even regeneration flagships. Examples of this are La Boqueria Market in Barcelona, Rotterdam's Market Hall, Borough Market in London, wet markets in Hong Kong or the Port Market in Montevideo. They are 'must visit' locations for international travellers looking for something different and authentic. But these transformations are clashing with markets' important role as public meeting places and ordinary everyday life places for the most vulnerable. The confluence of these potentially contradictory trends and processes turns markets into 'contested spaces'.

Throughout this book we will show markets as important spaces in the complex fabric and rhythm of our cities today. Markets come and go; sometimes they are fixed, sometimes they are itinerant; they can be regulated or informal. Traders and customers might be displaced from central cities only to re-emerge somewhere else. Sometimes markets are redeveloped and reimagined as gentrified or touristified spaces. Other times they are part of a resistance against this very same gentrification; and they can create opportunities for fairer forms of consumption in the face of unethical practices by multi-national corporations. We therefore conceptualise markets as spaces of contestation between different practices and discourses of city-making. What is clear is that markets are always evolving, changing, in flux and transformation.

There is growing interest in cities as global spaces for the contestation over processes of urban neoliberalisation, the effects of gentrification and inequalities, and also as incubators of just and fairer alternative ways of living (Harvey, 2012). Our book extends in an original way the analysis of these issues to retail spaces, which are generally marginalised within critical urban studies. In particular, we look at the traditional indoor or street markets in cities across the global North and South as new spaces for contestation.

It is impossible to define markets. There is an immense variety of markets throughout the world and across history and there is no comprehensive study of them, as no discipline has particularly analysed markets. In this book we discuss covered and indoor markets, some owned and managed by the state but also those which are managed and owned privately. We also discuss street markets and touch on informal and unregulated markets. Our aim is not to provide a typology or a fixed definition of markets, but to let the variety and geographical variegation flow and enrich our analysis. Seale (2016, p. 12) in her introduction to the book *Markets, Places, Cities* conceptualises markets as nodes, 'where material and intangible flows – of people, goods, times, senses, affect – come to rest, terminate, emerge, merge, mutate and/or merely pass through, and are contingent and relational to each other'. Following this view, we find it useful to think of markets in terms of social relations rather than fixed or bounded spaces. Indeed, if there is one common element to all different types and forms of markets is that they bring people together; they connect them in relationships of economic and extra-economic exchange. Reflecting this diversity and fluidity we do not use a uniform terminology across the book when referring to markets; some chapters use 'public markets', others 'traditional markets' and others use vernacular terms such as *tianguis* for the outdoor stalls in Mexico.

Street or informal markets and covered, municipal, indoor markets have not often been studied together. Street and informal markets and vendors have been studied profusely, particularly from an economic anthropology and urban informality studies perspective (Bhowmik, 2012; Bostic et al., 2016; Cross and Morales, 2007; Graaff and Ha, 2015; Seale and Evers, 2014). In contrast, research on covered and regulated markets is sparse and spread out across disciplines and geographies. In this book we look at them together. Most of our chapters look at covered and indoor market halls, but several chapters analyse the situation of street markets, though mainly formal and legal ones. Bringing all these kinds of markets together in one book is innovative in itself and our aim in gathering them is to show why and how, despite their variety, they can all become contested spaces in cities today.

This introductory chapter lays the theoretical foundations of the book which are then taken up, expanded and illustrated by the case study chapters. It provides a broad look at markets and ways of understanding their past and current role and situation. I offer a critical review of the global history of markets in order to help us situate their contemporary contested nature, particularly in relationship to notions of modernity. Following from this, I develop the three main analytical frameworks that will be broadly used by the rest of the case study chapters: markets as the frontier for processes of gentrification; markets as spaces for resistance and political mobilisation and markets as spaces for the development of alternative practices of production and consumption. Here I review various concepts and case studies drawn from existing literature gathered together from across various disciplines from retail studies to ethnography, geography, urban planning and sociology. The catalogue of theoretical and analytical concepts presented here will be picked up later by the case study chapters. In the final section of the introduction I present the structure of the book and briefly refer to our case study markets. All the markets discussed in this book are conceptualised as contested spaces but they do not all show the same features or processes. Therefore, not all case studies develop the three analytical frameworks equally; some are highlighted more than others, building and expanding on some concepts without necessarily developing others. As markets are incredibly variable and diverse, so is their analysis.

The book draws on original work carried out by researchers in the international network 'Contested Cities', which received funding from the European Union between 2012 and 2016 to explore processes of urban neoliberalisation and resistance in European and Latin American cities. Researchers from the network were based in the UK, Spain, Argentina, Chile, Brazil, Ecuador and Mexico. As explained in more detail in the Preface of this book, the network developed a sub-group to research the transformation of traditional markets that we were all witnessing in our cities and which we report on in this book.

Markets in the modern city narrative

In this section I provide a brief international history of markets, directly linking their transformations to ideas, notions and narratives of modernity. Through this short history we will see that markets have been, and continue to be, contested spaces because they are linked to disputed notions of modernity, formality and informality that continue until today. Cross (2000) has already advanced this discussion by exploring the changes in public discourse and policy on informal street vending in relation to the notions of modernity and postmodernity, and we are now extending some of these arguments to regular and formal markets.

Markets were, of course, key spaces in the pre-modern and the pre-colonial city as places for the exchange of produce and where the urban and the rural met. Markets both fed cities and were places of encounter between many different ethnicities and cultures. Stobart and Van Damme (2016) show how urban historiography has dedicated a great deal of attention to the role of marketplaces in European medieval history; markets made cities. Beyond Europe, in pre-colonial and pre-industrial cities, markets were also central as points of encounter in long-distance trade networks, where a variety of produce and ethnicities in customers and traders could be seen (Beeckmans and Bigon, 2016; Bromley et al., 1975; Guyer, 2015).

The onset of modernity and ideas of planning and building cities in a rational way interfered with how small retail was organised in cities. Two main processes seem to have affected markets. On the one hand, to make cities more efficient and aesthetically pleasing from a bourgeois perspective, informal and apparently disorganised and messy trade had to be tamed and tidied up (see Guàrdia and Oyón, 2015, for markets in Europe). According to Cross (2000, p. 40) 'street vending thus came under savage attack throughout the modernist era'. Shopkeepers and established formal traders also lobbied to eradicate this unfair competition, as street traders did not always pay taxes (Schmiechen and Carls, 1999). On the other hand, from the mid-nineteenth century, the state became more involved in food provisioning, alongside other public functions, in response to the rapid industrialisation and urbanisation of big cities (Guàrdia and Oyón, 2015) and there was a programme of building municipal market halls in many cities around the world. These two related processes, which I elaborate below, challenged informal street trading and aimed to establish carefully-regulated markets as commercial spaces for the middle and upper classes. These processes, however, were never complete, as we will see.

Modern efforts to tame and regularise markets

In London throughout the nineteenth century, informal street markets were abundant and fed the rapidly-growing population, especially the poorest (Kelley, 2016). However, they were regarded by the authorities as insanitary and traders were often persecuted and displaced to the peripheries (Jones, 2016). In the UK, the removal of informal markets was part of Victorian urban improvement initiatives (Jones, 2016). Covered market halls were built as an educational and moral tool to reform 'the most offensive kind of lower class street culture and people [...] who often made the old market their home' (Schmiechen and Carls, 1999, p. 55). The policy did not always work and, in London, informal street traders did not necessarily move in these new market halls as they found them too expensive and inconvenient for their customer base (Jones, 2016). The dream of the middle class and sanitised public market hall also failed as middle and upper class consumers favoured the newly-developed department stores such as Harrods or Selfridges, and by 1890s the 'market hall had become a working class department store' (Schmiechen and Carls, 1999, p. 193). Similarly, in Paris by the mid-1800s street vendors and street markets were also becoming a concern for the state. In particular, the central markets of Les Halles became a target as an insanitary and crowded space inundated by street traders not suitable for a modern imperial city (Thompson, 1997). Their redevelopment became part of the 1850s 'Haussmannisation' project to bring what was until then a dangerous, unruly and filthy space into the realm of the bourgeois city, not without resistance and protests from the traders (ibid).

In the global South, the relationship between markets, urbanisation and modernisation has parallels but also many differences. The building of markets in colonial cities was part of the 'urban civilisation' project of colonial powers. Markets were an important part of the rational plans for cities but, as shown above, modernity could not obliterate their informality and messiness. Beeckmans and Bigon (2016) show how markets built in Dakar and Kinshasa by French and Belgian authorities in late-nineteenth and early twentieth centuries were very cosmopolitan, which later became a concern for the colonial authorities worried about the sanitary consequences of ethnic mixing. In Latin America, street trading has always been part of street life from precolonial times, often with a conflictual relationship with state authorities (Bromley, 1998). In Rio de Janeiro, the first attempts to impose some regularity and symmetry onto a sprawling street trade goes back the colonial Portuguese authorities in 1789, supported by the upper and middle classes and influenced by hygienist discourses (Antunes Maciel and Leandro de Souza, 2012). By 1841, La Candelária covered market had been built to clean up some of this street trade; However, with the rapid growth of the city this area around the market soon became again a meeting place for the poorest in the city. Later in 1907 a new market was opened, hoping again to cleanse the area of undesirables (Antunes Maciel and Leandro de Souza, 2012, p. 67).

The building of the modern Western and colonial city was supposed to eradicate primitive forms of trade by bringing them into regularised markets, which would become exemplars of clean and orderly behaviour. The reality, of course, is that poor people did not go away and still needed cheap access to food and easy business start-ups provided by informal street markets. These kinds of markets, therefore, persisted in many cities of the global North well into the nineteenth century, and of course are still very much present in many other cities particularly in the global South. Indoor and regulated markets, although initially established in contrast to informal markets eventually acquired many of the same characteristics and stigmatisation as they attracted the poorest in cities. What we see is a continual effort by the modern state to tame not only street and informal trading but also any formal regular and indoor markets which challenge the modern values of order, standardisation, separation of functions and hygiene.

The rise and fall of the municipal markets as public assets

With the rapid urbanisation and industrialisation of cities through the nineteenth century, feeding the rising population became a major concern of the state, which was becoming by now a central form of organising modern society. In many countries, one way of dealing with this issue was to rein in markets under state control and turn them into key public services. The regulation of food provision and distribution became one of the most important roles of the incipient local governments in Europe. In France, the French Revolution saw the abolition of all feudal rights related to markets and these became now part of the municipal responsibility like hospitals, schools or jails (Guàrdia and Oyón, 2007). The role of these municipal markets in the spatial configuration of the city differed across countries (Guàrdia and Oyón, 2015). In the UK, iron-wrought central indoor markets were built at the heart of cities, particularly in northern England, although London continued the tradition of street markets (Schmiechen and Carls, 1999). This was followed to some extent in the Scandinavian countries although with less development of the municipal market. In contrast, a more polycentric model was adopted in Paris and then followed in Barcelona, Madrid, Turin or Berlin, cities with higher urban density where neighbourhood markets were built to serve the local population of the district (Fava, Guàrdia and Oyón, 2016). In Mexico City, a city which features prominently in our book, local authorities started to get more involved in the regulation of markets from the beginning of the nineteenth century and by 1850s the first municipal covered market was built (González, 2016). In the USA a tightly-regulated system of municipal food provision was developed through a network of public markets from the eighteenth to mid-nineteenth century. In New York, for example, there was a political consensus that the municipal model served to ensure the public good of citizens' access to food (Baics, 2016). But by around the 1820s, the council was struggling to keep up their investment in public markets in relation to urban growth; This coincided with trends to deregulate food provisioning at a national level and from then on the idea of the public market entered in a decline (Baics, 2016).

This decline of the municipal market model in New York was mirrored in Europe, and by the beginning of the twentieth century the model was in crisis. The UK, which had been a pioneer of the covered market hall, was the first to move to a new food retailing system with large wholesalers and food chains, according to Guàrdia and Oyón (2015). These two authors mark the 'definitive crisis' of the covered municipal market after the Second World War with a convergence of several factors: destruction of the physical infrastructure, individual car ownership, the dispersion of the population and the supermarket revolution. We can add to these factors the shift in the role that local authorities would play, particularly from the 1970s, as they moved from being predominantly managers and deliverers of public services to being urban entrepreneurs as well (Harvey, 1989). Local authorities, particularly in the North Atlantic zone, became increasingly under pressure to generate income; central government funds would start to withdraw, so cities had to attract and compete for external private investment (Brenner and Theodore, 2002). In this new phase, public markets, often serving the poorest in cities were a hindrance to the neoliberalising spirit of many local leaders.

Narratives of decline, resurgence and commodification

As I have already suggested above, the modern narrative was for informal and street trading to disappear. The story went that market halls in Europe and in Latin America, the focus of our book, were the natural evolution that would formalise, dignify and impose order in small street retail. The next step with globalisation was for market halls to be superseded by supermarkets and similar multi-chain retailers. However, this modern narrative has not fully become reality (Cross, 2000). Street, informal trading and market and indoor halls are still very much present in cities and for many different reasons even growing; sometimes reflecting new tastes and desires for a more authentic type of consumption and sometimes due to the lack of other forms of employment.

This narrative of the decline and obsolescence of markets, a common thread in our book, has partly been promoted by the academic discourse. Stobart and Van Damme show that while marketplaces occupy the interest of medieval historians, this focus disappears when it comes to 'narrative accounts on urban modernization in the industrial age' (2016, p. 361). The emphasis for retail historians has been on analysing innovations, prioritising the novelty of the department store, chains, internationalisation etc. And in terms of contemporary research, there is little research on markets from a retail perspective. Retail theory has generally been dominated by neoclassical assumptions and evolutionary ideas, such as the 'retail wheel', which understands that new forms of retailing start as low-cost and with low margins, and then naturally evolve and expand by 'trading up', searching for wealthier customers and selling at higher prices (Brown, 1995). Of course, markets as we approach them in this book do not always fit this 'retail wheel' and can often remain stubbornly at its beginning. As Brown (1995) notes, these theories are based on universal assumptions of linear evolution, mostly in retail environments of the global North. These dominant modernist narratives have therefore overlooked and underestimated the role of markets, generally relegating them to a marginal place in our cities, as relics of past times. From a different perspective, research interest in markets has come from ethnographers and anthropologists (and occasionally urban sociologists and geographers) interested in informality and generally approaching them from a 'development' angle and hence most of times also marginalised from the main circuits of academic knowledge production.

At the same time as this marginalisation and decline, many of our case studies in the book show that markets are also experiencing a resurgence and many are being redeveloped (Seale, 2016) and rebranded to cater for an increasing demand for a niche shopping experience related to local food, authenticity and as a turn away from the omnipresence of supermarkets. This recent resurgence, in fact, repositions markets into their pre-modern role, but in a way which is commodified and ready for consumption; in a postmodern society, saturated with exchange value, the market becomes a place where we can temporarily seek a break and explore the opportunity to consume 'tradition'.

Contested markets: Three analytical frameworks

Bringing this historical analysis up to the contemporary phase, our book presents an innovative way of looking at markets as contested spaces from three overlapping perspectives, which we use as threads for our case studies. These three analytical angles are complex and intermesh with multiple processes that have been studied by a variety of academic disciplines. In the following pages we sketch out these processes, laying out a mosaic of conceptual tools and concepts which are then taken up, expanded and fleshed out by the case study chapters in the book.

Markets as frontier spaces for processes of gentrification

Our first analytical perspective looks at markets as contested spaces which are undergoing processes of gentrification. Gentrification is a concept often used to describe and criticise changes in the social make-up of a neighbourhood (Clark, 2005; Lees et al., 2013). Simply put, it can be seen as the replacement of a working class and/or low income population by middle class as dwellers, consumers or both. It is a complex process involving changes in the built environment, services (such as education) and the retail mix which shifts to accommodate a wealthier population. Most research into this process focuses on the residential urban environment and on the residents' (old and new) experiences. There has been much less research on how gentrification also takes place through and affects the retail environment and how users (old and new) are also affected. A definition we have used before defines retail gentrification in these terms:

> the process whereby the commerce that serves (amongst others) a popula-tion of low income is transformed/replaced into/by a type of retail targeted at wealthier people. From a different angle, we can also see it as the increase in commercial rents that pushes traders to increase price of their products, change products or change location.
>
> (González and Dawson, 2015, p. 19)

Within the contours of this broad definition, the gentrification of markets can take many different forms, as we will see in our case studies. Sometimes it can be part of a wider residential neighbourhood gentrification process, where the retail offer is 'upgraded' to fit the new residents while the older ones are priced out or left out of place. Other times markets can be part of 'retail-led' regeneration strategies, where the market itself could become a flagship for the redevelopment of an area. In historical city centres, like those in Latin America, the gentrification of markets often forms part of 'heritage-led' redevelopment strategies; buildings are restored and upgraded and in the process the old occupants and their clientele are partly or totally replaced.

These processes of gentrification are related to the recent resurgence of public markets discussed above in which they are re-discovered and consumed as a traditional experience. Markets are reframed as 'authentic' pre-modern shopping experiences, where users escape the anonymity and antiseptic environment of the supermarket. Zukin (2009) has already highlighted the importance of shopping for authenticity as a driver for gentrification. High-income residents or tourists, in the face of standardised suburban or hyper-corporate city centres, crave a sense of place, a return to what gave cities and neighbourhoods an identity and distinctiveness. This can be a particular typology and look of housing and a distinct retail landscape: independent coffee shops, small DIY shops, grocery stores or a traditional bakery. These places develop a marker and an aesthetic of authenticity which has been associated with the hipster sub-culture (Hubbard, 2016). Significantly, this process of 'boutiquing', as identified by Zukin et al (2009, p. 48) 'mark[s] an area as safe for commercial investment that will upgrade services and raise rents', leading to further rounds of retail and residential gentrification. Consequently, the original stores often cannot afford the new rents and have to abandon their businesses; sometimes, they might adapt them to reflect the new customer base, for example changing their products. Either of these processes end up displacing the previous users of the neighbourhood retail offer, as they can no longer find or afford the products they used to consume, or they simply feel excluded from this upscale and often racially-marked environment (Sullivan and Shaw, 2011). Retail gentrification is happening in many places and can take many forms, and some work on it is emerging (for example, Bridge and Dowling, 2001; González Martínez, 2016; Pascual-Molinas and Ribera-Fumaz, 2009; Schlack and Turnbull, 2015; Sequera and Janoschka, 2015).

As we will see later in the book, in many places processes of gentrification are directly linked to previous phases of abandonment or disinvestment. As I explained above, in many countries, markets as municipal services have suffered a decline which has been aggravated by the emergence and massive expansion globally of supermarkets and retail chains. We have argued before (González and Waley, 2013) that this situation of disinvestment pushes public markets into the gentrification frontier in a similar way in which Neil Smith (1996) analysed inner-city and working class neighbourhoods. Because of their traditional role as meeting and trading places, markets are very often in central and strategic urban locations. In the context described above of middle class search for authentic experiences and the neoliberalisation of urban policies, state authorities, who often own markets, are under great pressure to upgrade them and turn them into more profitable assets. As illustrated in our case studies, this retail upgrading leads to the displacement not only of the older traders but also of market customers and users and potentially of surrounding independent businesses.

There is a wealth of research documenting the displacement of street and informal traders, particularly in cities in the global South. Bromley and Mackie (2009) make an explicit connection between the process of gentrification in central areas in Latin American cities and the efforts by state actors to remove street traders from them. The process of displacement is generally preceded by a campaign to stigmatise of traders. In a similar way to how street traders were described in London or Paris in the nineteenth century, today 'street vendors are described like locusts, coming in "plagues," "droves," and "deluges," and the city is depicted as being both invaded and asphyxiated' (Bromley, 2000. p. 10). In our book, we show how these processes of displacement are also occurring in formally-regulated markets, not only in streets but also in covered market halls in cities in the global North. These displacement processes are, however, usually less dramatic and aggressive than the eviction of street vendors and therefore we need more nuanced ways to understand gentrification-induced displacement, such as those discussed by Marcuse (1985) and Slater (2009). This includes paying attention not only to the displacement of traders but also of those who use the market for shopping and socialising as well as the displacement of products and practices.

Markets as spaces for mobilisation, contestation and debate over public space and the city

Our second analytical framework looks at markets as spaces for political mobilisation, contestation and struggle for public space. As already stressed, markets form part of wider processes of urban and retail change, involving a diversity of actors which often turn them into spaces for contestation. Projects for the refurbishment, redevelopment or modernisation of markets frequently do not progress quietly or in an uncontested manner. On the contrary, market transformations are often challenged by the traders themselves, neighbours, customers and other adjacent traders and urban actors. Authorities at different state levels and private developers can have disparate views about the present and future of markets, and this brings them into potential conflict. Because of the central role that markets and street trading have historically played in cities and their public nature, contestation around markets often spills into much wider debates about the city, and in particular the right to the city for the most vulnerable.

There already exists an abundant literature on informal trading and con-testation for public space, particularly in cities of the global South. In Latin America, the work of Bromley and Mackie (2009) reports on the resistance of informal traders against their displacement from the centre of Cusco in Peru. Apart from planned demonstrations, traders also occupied or re-occupied the streets they were being displaced from, or nearby less surveilled areas. Crossa (2009) has analysed the complex practices of street traders as they struggle to remain in the historic centre of Mexico City. Most of the time, conflicts see street traders confronting state authorities, but struggles also emerge between formal and informal traders (see Bromley and Mackie, 2009; Omoegun, 2015).

These forms of political mobilisation by informal and street traders can be regarded as part of a labour struggle. Street traders around the world, who most of the time belong to marginalised groups, share workplace insecurity, harassment, precarity and eviction from their workplace (Graff and Ha, 2015; Habitat, 2015; Roever and Skinner, 2016) and these are the main issues they mobilise around. But often the contestation around markets goes beyond labour issues to become a struggle for the right to use the urban space to develop a livelihood. Central urban spaces attract the greatest footfall from residents, worker-commuters and tourists, and informal traders want to take advantage of this possibility for trade. But, as discussed already, state authorities increasingly implement aggressive techniques to 'recover' the streets for safe and sanctioned activities in the public space. Streets, particularly in populous global South cities, become a veritable battleground between authorities and informal traders (Mackie et al., 2014).

Informal trader mobilisations, however do not only concern these spatial battles for streets in order to develop their livelihoods, but have enlarged to make wider claims about the right to the city and debates about the use of public spaces. These claims have been strongly developed by national and international networks of informal workers and traders. For example, Streetnet, which emerged in Durban in 2002, launched a campaign in 2006 in alliance with other organisations representing the urban poor in response to what was seen as an exclusive view of the city particularly in the wake of the football World Cup in South Africa in 2010 (Streetnet, 2010). In the buildup to Habitat 3, the UN Summit on urban issues held in Quito in October 2016, informal and street trader organisations and networks mobilised to make sure their demands for an inclusive city would be listened to (WIEGO, n.d.). Informal traders, then, are not only reactive or defensive actors narrowly concerned with their trading space but often develop an important 'urban voice' (Brown et al., 2010) which can have an impact in planning and urban policy decision-making.

The overwhelming majority of research on traders and markets as actors and spaces for urban mobilisation is focused on cities in the global South and around informal and/or street trading, with the dispute over public space and displacement being the major triggers for contestation. In this book, however, we make an original contribution by showing that these struggles are now increasingly affecting regular and formal markets, which brings a new dimension as we will see in our case studies. Furthermore, in this book we also analyse mobilisations not only by traders but also by market users and shoppers, which tend to mobilise against the disappearance or marginalisation of traditional retail markets, as a sign of the loss of local identity in the retail landscape and of retail gentrification. Broad alliances between traders, local community groups and campaigns can emerge to defend 'our market', as we will see in the following case studies. In these cases, the neighbourhood and/or the city are invoked as the scale of the struggle, making the markets a metaphor or a symbol for wider urban struggles.

Markets as spaces for building alternative and counter-practices of production and consumption

Our third main analytical framework is that of markets as spaces for building alternative and/or counter-practices of production and consumption outside or beyond corporate networks. We also consider within this sphere the ways in which markets can create spaces for solidarity and social inclusion creating what Mele et al. (2015, p. 104) refer to as 'corrective spaces' to the 'excesses of advanced urbanism' characterised by corporate consumerism, excessive regulation of public spaces and the erosion of indigenous urban landscapes. Therefore, we argue that some of the practices surrounding the use of this 'traditional' type of retail can be considered a form of resistance or a kind of counter-practice to processes of neoliberalisation.

One of the most well-recognised features of traditional retail markets is that they are not simply retail spaces for the monetary exchange of goods but they are also social spaces where traders and customers develop meaningful relationships. This social function is particularly important to the most vulnerable residents of cities such as the elderly, single mothers, young people, ethnic minorities, migrants and refugees and people with health problems (Morales, 2009; Project for Public Spaces, 2003; Watson and Studdert, 2006). These inclusive features ran counter to the kind of urbanism and aspirational cities that prevail with mega-developments, privatised and securitised spaces and increasingly unaffordable and precarious housing. Markets can therefore act as a kind of refuge for excluded groups of people.

Markets are also particularly important in creating spaces for social interaction between groups of people from diverse backgrounds, age, class, culture and ethnicity. Cities across the world are becoming more diverse but this does not necessarily mean that citizens necessarily interact with people from diverse backgrounds in their everyday life. It seems, however, that markets can open up spaces for interaction as they throw people together in a less conflictual way than in other spheres such as the housing or labour market. Watson (2009) observes how people in markets 'rub along', sharing the same spaces and although not necessarily engaging in deep interaction, at least tolerating each other, potentially creating practices of tolerance. Indeed, ethnographers have looked at markets as a microcosm of cosmopolitanism (Duruz, Luckman and Bishop, 2011) or what Anderson (2011) calls a 'cosmopolitan canopy'. We should not however fall into a romantic notion of the market and see them as havens for ideal inter-ethnic and inter-class cohabitation. Pardy (2005 cited by Hiebert et al., 2015, p. 11) argues that in fact markets might develop these inclusive features precisely because they could be places of 'indifference to difference' where people can practice 'mutual avoidance' and share a space without meaningful interaction.

Markets can also be spaces for solidarity and this can happen in many different forms. Even though markets are by definition spaces for the monetary exchange, our own experience and the literature shows that plenty of non-monetary exchanges happen in markets. For example, as shown in some of our case studies, market traders report on caring for customers, particularly the elderly, those feeling lonely or those with mental health issues. As reported by Stillerman and Sundt in the case of street markets in Santiago de Chile, 'personal networks and acts of reciprocity are significant in relationships among vendors and between vendors and customers' (2007, p. 192). This trust, they argue, is developed because of the absence of state supervision and the lack of formal agreements. Our research and observation in formal and regulated markets shows that these reciprocal relationships also take place there, admittedly where there is also an element of informality and a more direct relationship between customers and sellers than in shopping malls, for example.

Another form of solidarity economy is when markets and/or market traders actually use alternative circuits and production and distribution networks from the corporate and globalised ones as in the case study of Mercado Bonpland (Chapter 8). Within this broad field of solidarity economies, we find a big range of different types of markets, from the rekindled farmers' markets so popular in the US and UK to the traditional subsistence markets of small towns and settlements across the world (Bubinas, 2011). In the US, farmers' markets that connect producers with consumers directly have been used as tools for social mobility and community cohesion (Project for Public Spaces, 2003). But they also have been criticised for creating exclusionary spaces, as food can be significantly more expensive and they often run by and frequented by middle class white consumers (Alkon and McCullen, 2011).

There are therefore many ways in which markets can become alternative spaces to the commodification of social relations and segregation between groups that we increasingly experience in cities. Sometimes, using and working in markets can be conscious and political micro-acts of resistance, but perhaps more often they are subtle forms of 'subaltern urbanisms' (Roy, 2011) practiced by ordinary people in their everyday life.

Our case study markets

The case studies in our book represent the variety and diversity of markets in cities across Europe and Latin America. In particular, we discuss markets in Mexico City, Buenos Aires, Santiago de Chile, Quito, Madrid, London, Leeds and Sofia. Our markets, therefore, are all situated in relatively large cities ranging from the very large Mexico City (8.8m people) to Sofia (1.2m) and Leeds (750,000). As such, the markets are immersed in complex urban processes and transformations. As discussed above, the innovation of this book is that we mainly focus on formal and regulated markets as opposed to the majority of the academic literature in this area, which tends to focus on informal street markets. Most of our case studies are also covered and indoor markets, often owned by public authorities or in cooperation by the traders themselves. However, as we have started to develop in the introduction of this book, despite their formal, regulated and fixed nature, they are also somewhat informal spaces and they are, in the main, spaces at the margin of the official discourses of the city. In our case study markets we develop the three analytical frameworks discussed above to show how markets have become contested spaces in our cities.

The first market that we present in the book is that of La Merced, situated at the edge of the historic centre of Mexico City (Chapter 2). This is a complex of several indoor markets surrounded by formal and informal street traders with thousands of people selling an enormous variety of produce now under pressure for redevelopment. La Vega Central in Santiago de Chile (Chapter 3) is also a big complex of covered markets at the edge of the city centre, owned and managed by the collective of traders, which gives it an interesting political role. The market and the wider neighbourhood, historically a space for low-income and marginalised workers and customers, is now seeing signs of early gentrification and it remains to be seen how the traders might manage this process.

In London, we also discuss the gentrification and real estate speculative pressures on markets and we focus on three different trader and/or customer campaigns which have emerged to resist the demolition or transformation of their markets (Chapter 4). The theme of political organisation around markets is then taken up by another chapter (Chapter 5) on Mexico City, this time discussing the formally organised *tianguis* markets, made up of street stalls. The chapter discusses the struggle for the use of public space and the political role of the leaders of the trader associations.

In a comparative chapter between Madrid and Mexico City (Chapter 6), we analyse the process of 'gourmetisation' of markets, a trend related to the rise of a 'foodie' culture amongst middle class consumers looking for authentic experiences that they can find in markets branded and marketed as 'traditional'. Chapter 7 looks in more detail at the recent and uneven trans-formations affecting the relatively large network of municipal markets in Madrid, amidst a process of retail gentrification in many neighbourhoods of the city and the power of global retail chains. In Chapter 8 we visit Buenos Aires, and analyse the 'solidarity market' Bonpland, a small indoor market which was taken up by a neighbourhood assembly after the 2001 Argentinian economic crisis and which has become a hub for alternative consumption and production networks across the country.

The cases in our remaining cities in Leeds, Sofia and Quito (Chapters 9, 10 and 11) all show markets which, because of their historical marginalisation by the authorities, have become refuges for marginalised groups. However, this is now under threat because of processes of redevelopment. In Leeds, Kirkgate Market has become a culturally and ethnically superdiverse space, where people from very different backgrounds interact and develop networks of care and reciprocity, but it is unclear whether its current transformations and the local authority's aspirations for an upscaled market will guarantee this inclusivity. In Sofia, the so-called 'Women's Market' was for decades stigma-tised and connected to pre-modern notions of Eastern Europe. The chapter shows how the market is still a space for marginalised communities but not necessarily integrated, with groups interacting side-by-side. In Quito, the San Roque Market has become a place for the reproduction of the popular culture of the indigenous groups manifested in collective organising around festive and religious traditions, but there is a long struggle with the authorities who have repeatedly tried to move and dismantle the market and its diverse community.

In the chapters which follow, we develop these case studies in depth by building on the three analytical frameworks presented in this introduction. The diversity of markets themselves and their relationship to the neighbourhoods and cities where they are based as well as the variety of expertise of the researchers means that the case study chapters draw on the three analytical frameworks in differing ways. The concluding chapter returns to these analytical threads with a comparative analysis of the markets, reflecting on the similarities and difference across them.

References

Alkon, A.H. and McCullen, C.G. (2011). Whiteness and farmers markets: Performances, perpetuations … contestations? *Antipode*, 43(4), 937–959.

Anderson, E. (2011). *The Cosmopolitan Canopy: Race and Civility in Everyday Life.* New York: W.W. Norton.

Antunes Maciel, L. and Leandro de Souza, V. (2012). Ordem na praça: Normas e exercício de administração em mercados do Rio de Janeiro. *Revista Internacional de História Política e Cultura Jurídica*, 4(1), 55–80.

Baics, G. (2016). The geography of urban food retail: Locational principles of public market provisioning in New York City, 1790–1860. *Urban History*, 43(3), 435–453.

Beeckmans, L. and Bigon, L. (2016). The making of the central markets of Dakar and Kinshasa: From colonial origins to the post-colonial period. *Urban History*, 43(3), 412–434.

Bhowmik, S. (2012). *Street Vendors in the Global Urban Economy.* London: Taylor & Francis.

Bostic, R.W., Kim, A.M. and Valenzuela Jr, A. (2016). Contesting the streets: Vending and public space in global cities. *Cityscape*, 18(1), 3–10.

Brenner, N. and Theodore, N. (2002). Cities and the geographies of "actually existing neoliberalism". *Antipode*, 34(3), 349–379.

Bridge, G. and Dowling, R. (2001). Microgeographies of retailing and gentrification. *Australian Geographer*, 32(1), 93–107.

Bromley, R. (2000). Street vending and public policy: A global review. *International Journal of Sociology and Social Policy*, 20(1/2), 1–28.

Bromley, R.D. and Mackie, P.K. (2009). Displacement and the new spaces for informal trade in the Latin American city centre. *Urban studies*, 46(7), 1485–1506.

Bromley, R.D. (1998). Market-place trading and the transformation of retail space in the expanding Latin American city. *Urban Studies*, 35(8), 1311–1333.

Bromley, R.J., Symanski, R. and Good, C.M. (1975). The rationale of periodic markets. *Annals of the Association of American Geographers*, 65(4), 530–537.

Brown, A., Lyons, M. and Dankoco, I. (2010). Street traders and the emerging spaces for urban voice and citizenship in African cities. *Urban Studies*, 47(3), 666–683.

Brown, S. (1995). Postmodernism, the wheel of retailing and will to power. *The International Review of Retail, Distribution and Consumer Research*, 5(3), 387–414.

Bubinas, K. (2011). Farmers markets in the post-industrial city. *City & Society*, 23(2), 154–172.

Clark, E. (2005). The order and simplicity of gentrification: a political challenge. In: Rowland, A. and Bridge, G. (Eds.), *Gentrification in a Global Context: The New Urban Colonialism.* London: Routledge, pp. 261–269.

Cross, J. (2000). Street vendors, and postmodernity: Conflict and compromise in the global economy. *International Journal of Sociology and Social Policy*, 20(1/2), 29–51.

Cross, J. and Morales, A. (Eds.) (2007). *Street Entrepreneurs: People, Place, & Politics in Local and Global Perspective*. London: Routledge.

Crossa, V. (2009). Resisting the entrepreneurial city: Street vendors' struggle in Mexico City's historic center. *International Journal of Urban and Regional Research*, 33(1), 43–63.

Duruz, J., Luckman, S. and Bishop, P. (2011). Bazaar encounters: Food, markets, belonging and citizenship in the cosmopolitan city. *Continuum: Journal of Media & Cultural Studies*, 25(5), 599–604.

Fava, N., Guàrdia, M. and Oyón, J.L. (2016). Barcelona food retailing and public markets, 1876–1936. *Urban History*, 43(3), 454–475.

González, G.M. (2016). La gestión municipal: ¿Cómo administrar las plazas y los mercados de la ciudad de México? 1824–1840. *Secuencia*, 95(May/August), 39–62.

González, S. and Waley, P. (2013). Traditional retail markets: The new gentrification frontier? *Antipode*, 45(4), 965–983.

González, S. and Dawson, G. (2015). Traditional markets under threat: Why it's happening and what traders and customers can do. Available from: http://eprints.white rose.ac.uk/102291/

González Martínez, P.G. (2016). Authenticity as a challenge in the transformation of Beijing's urban heritage: The commercial gentrification of the Guozijian historic area. *Cities*, 59, 48–56.

Graaff, K. and Ha, N. (2015). Introduction. In: Graaff, K. and Ha, N. (Eds.), *Street Vending in the Neoliberal City: A Global Perspective on the Practices and Policies of a Marginalized Economy*. Oxford: Berghahn Books, pp. 1–15.

Guàrdia, M. and Oyón, J.L. (2007). Los mercados públicos en la ciudad contemporánea. El caso de Barcelona. *Biblio 3W, Revista Bibliográfica de Geografía y Ciencias Sociales*, Universidad de Barcelona, Vol. XII, n° 744. Available from: http://www.ub.es/geocrit/b3w-744.htm

Guàrdia, M. and Oyón, J. (2015). Introduction: European markets as markets of cities. In: Guàrdia, M. and Oyón, J. (Eds.), *Making Cities Through Market Halls. Europe, 19th and 20th Centuries*. Barcelona: Museu d'Història de Barcelona, Institut de Cultura, Ajuntament de Barcelona, pp. 11–71.

Guyer, J. (2015). Markets and urban provisioning. In: Monga, C. and Lin, J.Y. (Eds.), *The Oxford Handbook of Africa and Economics*. Volume 1. Oxford: Oxford University Press, pp. 104–114.

Habitat (2015). Informal sector. *Habitat III Issue Paper 14*. New York: Habitat. Available from: http://unhabitat.org/wp-content/uploads/2015/04/Habitat-III-Issue-Paper-14_Informal-Sector-2.0.pdf

Harvey, D. (1989). Capitalism from managerialism to entrepreneurialism in urban governance transformation. *Geografiska Annaler*, 71(1), 3–17.

Harvey, D. (2012). *Rebel Cities: From the Right to the City to the Urban Revolution*. London: Verso Books.

Hiebert, D., Rath, J. and Vertovec, S. (2015). Urban markets and diversity: Towards a research agenda. *Ethnic and Racial Studies*, 38(1), 5–21.

Hubbard, P. (2016). Hipsters on our high streets: Consuming the gentrification frontier. *Sociological Research Online*, 21(3), 1. Available from: http://www.socresonline.org.uk/21/3/1.html

Jones, P.T. (2016). Redressing reform narratives: Victorian London's street markets and the informal supply lines of urban modernity. *The London Journal*, 41(1), 60–81.

Kelley, V. (2016). The streets for the people: London's street markets 1850–1939. *Urban History*, 43(3), 391–411.

Lees, L., Slater, T. and Wyly, E., (2013). *Gentrification*. Abingdon, UK: Routledge.

Mackie, P.K., Bromley, R.D.F. and Brown, A. (2014). Informal traders and the battlegrounds of revanchism in Cusco, Peru. *International Journal of Urban and Regional Research*, 38(5), 1884–1903.

Marcuse, P. (1985). Gentrification, abandonment, and displacement: Connections, causes, and policy responses in New York City. *Washington University Journal of Urban and Contemporary Law*, 28, 195–240.

Mele, C., Ng, M. and Chim, M.B. (2015). Urban markets as a 'corrective' to advanced urbanism: The social space of wet markets in contemporary Singapore. *Urban Studies*, 52(1), 103–120.

Morales, A. (2009). Public markets as community development tools . *Journal of Planning Education and Research*, 28(4), 426–440.

Omoegun, A. (2015). Street trader displacements and the relevance of the Right to the City concept in a rapidly urbanising African city: Lagos, Nigeria. PhD thesis, School of Planning and Geography, Cardiff University, UK.

Pascual-Molinas, N. and Ribera-Fumaz, R. (2009). Retail gentrification in Ciutat Vella, Barcelona. In: Porter, L. and Shaw, K. (Eds.), *Whose Urban Renaissance? An International Comparison of Urban Regeneration Strategies*. London: Routledge, pp. 180–190.

Project for Public Spaces (2003). *Public Markets as a Vehicle for Social Integration and Upward Mobility*. New York: The Ford Foundation. Available from: http://www.pps.org/pdf/Ford_Report.pdf

Roever, S. and Skinner, C. (2016). Street vendors and cities. *Environment and Urbanization*, 28(2), 359–374.

Roy, A. (2011). Slumdog cities: Rethinking subaltern urbanism. *International Journal of Urban and Regional Research*, 35(2), 223–238.

Schlack, E. and Turnbull, N. (2015). Emerging retail gentrification in Santiago de Chile: The case of Italia-Caupolicán. In: Lees, L., Shin, H.B. and López-Morales, E. (Eds.), *Global Gentrifications: Uneven Development and Displacement*. Bristol: Policy Press, pp. 349–373.

Schmiechen, J. and Carls, K. (1999). *The British Market Hall: A Social and Architectural History*. New Haven, CT: Yale University Press.

Seale, K. (2016). *Markets, Places, Cities*. London: Routledge.

Seale, K. and Evers, C. (2014). Informal urban street markets: International perspectives. In: Seale, K. and Evers, C. (Eds.), *Informal Urban Street Markets*. London: Routledge, pp. 1–14.

Sequera, J. and Janoschka, M. (2015). Gentrification dispositifs in the historic centre of Madrid: A re-consideration of urban governmentality and state-led urban reconfiguration. In Lees, L., Shin, H.B. and López-Morales, E. (Eds.), *Global Gentrifications: Uneven Development and Displacement*. Bristol: Policy Press, pp. 375–394.

Slater, T. (2009). Missing Marcuse: On gentrification and displacement. *City*, 13(2–3), 292–311.

Smith, N. (1996). *The New Urban Frontier: Gentrification and the Revanchist City*. London: Routledge.

Stillerman, J. and Sundt, C. (2007). Embeddedness and business strategies among Santiago, Chile's street and flea market vendors. In: Cross, J. and Morales, A. (Eds.),

Street Entrepreneurs: People, Place, & Politics in Local and Global Perspective. London: Routledge, pp. 180–200.

Stobart, J. and Van Damme, I. (2016). Introduction: Markets in modernization: transformations in urban market space and practice, c. 1800–c. 1970. *Urban History,* 43(3), 358–371.

Streetnet (2010). Campaigns work report to Streetnet Congress August 2010 World Class Cities for All Campaign (WCCA), Durban, South Africa. Available from: http://www.streetnet.org.za/docs/reports/2010/en/WCCA-Campaign-Report-August-2010.pdf

Sullivan, D.M. and Shaw, S.C. (2011). Retail gentrification and race: The case of Alberta Street in Portland, Oregon. *Urban Affairs Review,* 47(3), 413–432.

Thompson, V. (1997). Urban renovation, moral regeneration: Domesticating the Halles in second-empire Paris. *French Historical Studies,* 20(1), 87–109.

Watson, S. and Studdert, D. (2006). Markets as sites of social interaction. Spaces of diversity. York: The Joseph Rowntree Foundation. Available from: https://www.jrf.org.uk/sites/default/files/jrf/migrated/files/1940-markets-social-interaction.pdf

Watson, S. (2009). The magic of the marketplace: Sociality in a neglected public space. *Urban Studies,* 46(8), 1577–1591.

WIEGO (n.d.). Habitat III. Women in informal employment: Globalising and organising [website]. Available from http://wiego.org/cities/habitat-iii

Zukin, S., (2009). *Naked City: The Death and Life of Authentic Urban Places.* Oxford, UK: Oxford University Press.

Zukin, S., Trujillo, V., Frase, P., Jackson, D., Recuber, T. and Walker, A., (2009). New retail capital and neighborhood change: Boutiques and gentrification in New York City. *City & Community,* 8(1), 47–64.

2 Markets of La Merced

New frontiers of gentrification in the historic centre of Mexico City[1]

Victor Delgadillo

Introduction

This chapter analyses the dispute over the markets of La Merced in the context of public policies for the modernisation of food markets and the 'rescue' of urban heritage. After a fire in the largest market in February 2013, the Government of the Federal District (GDF) organised an urban-architectural competition for the 'comprehensive rescue'[2] of La Merced. The area of La Merced is part of the historic centre of Mexico City, accommodating a large quantity of formal and informal trading in eight public markets, various shopping centres, warehouses, shops and hundreds of street vendors. The government intends to seize the 'unpostponable opportunity' to explore the 'possibilities' of modernising La Merced. The megaproject, now postponed, aims to 'recover' the area through the creation of public space, redevelopment of the markets, an increase in building density, improved mobility, modernisation of the business model and the creation of a tourist centre for gastronomy.

The megaproject is part of a trend adopted by local and national governments to gradually abandon domestic production and traditional systems of food distribution to the detriment of food sovereignty, whilst at the same time promoting food imports and increasing the presence of domestic and foreign supermarket chains. According to these governments, the traditional supply channels are inefficient, ineffective and an obstacle to 'modernisation', while public markets are physically deteriorated, functionally and economically obsolete and should be renewed and modernised (SEDECO, 2015).

This chapter is organised as follows: 1. A brief review of theoretical contributions to the concept of 'urban frontiers' in order to emphasise the porous but clear barriers that separate the 'modern city' from the 'traditional city of the markets'. 2. A discussion of the 'regeneration' model of the historic centre driven by the local government in alliance with certain private investors and which since 2009 has spread to other neighbourhoods. 3. A presentation of the main disputes over public markets in the megacity of Mexico City, marked by a discourse of obsolescence and deterioration. 4. An analysis of the megaproject and the dispute over the markets of La Merced along with the interests and views of the main stakeholders. This chapter is based on a study of the urban heritage of the neighbourhood and the markets of La Merced (Delgadillo, 2014), involving qualitative and quantitative methods of analysis including public statistics, field observations, interviews, ethnographic research and surveys. Analysis of the megaproject and the problems of markets is undertaken by analysing critically public policies. We understand public policies as fundamentally political because they are the sphere for discussion and contestation between political parties and groups for different visions of development in the broadest sense of the term. Public policies are not only defined between leaders of different hierarchies and large and small investors, but in some situations civil society plays an important role in disputes over policies and public assets.

Urban frontiers

The frontier is an intrinsic concept for cities. Cities are social and historical products with physical and virtual boundaries that demarcate rural–urban boundaries and separate different political and administrative entities with different powers over the territory. The city, an environment built by generations of people for better living and for integration and relation with others, is a circumscribed, finite and bounded territory that responds to the culture of the limits (Mongin, 2006). However, this city, a symbol of emancipation and social integration, is now confronted with a metropolitan dynamic and globalisation that divides, scatters, fragments, privatises, decentralises, separates and creates new and diverse urban and territorial hierarchies.

Cities have internal frontiers that separate the private from the public and define appropriate territories for different social groups. The frontier is a concept that can make conclusive and impassable dividing lines, but also barriers that are porous, diffuse and extremely complex. The frontiers are physical, symbolic, tangible, intangible, artificial or natural. Frontiers are social constructions in time for: demarcating social appropriation; separating places and social groups; demarcating identities and belonging; and defining spatial, physical and symbolic domains (Rapoport, 1972). The (im)permeable frontiers mark the contours of socio-spatial segregation, while stronger borders accentuate the limits of insular urbanism and the fragmented city: walls, fences and gates that accompany video surveillance, police or private security.

In his study of gentrification, Smith (2012) addressed the expansion of the frontiers of a revanchist urbanism for which the poor, who have dared to live in the centre, have had to pay for. Smith makes an analogy between the conquest of the Wild West of the United States and the young conquerors from the late twentieth century, the 'urban cowboys' who are introduced into neighbourhoods considered deteriorated and dangerous in some US cities, to tame, domesticate and (re)colonise them. For Smith, more than the 'urban cowboys' (or 'pioneers' for other authors) it was the two industries of real estate and culture that redefined the new 'urban frontier'. The first is a voracious industry that pursues profit; while the second, an ally of the former, reproduces the dominant culture, including its version of 'counterculture' to attract tourists, consumers, public, sponsors and investors, which together promote gentrification.

These concepts of frontier and urban revanchism are of interest in understanding the urban transformations in Mexico City and in particular in the area of La Merced. Without specifically talking about frontiers, Delgadillo (2005) and Monnet (1995) nevertheless recognise the historical bipartite condition of the historic centre of Mexico City. This is characterised by the better physical conditions of the southern and western areas with a lower population density, where the 'modern' and trading functions and services for the middle classes are located (sometimes referred to as 'The City of Palaces') and the areas to the north and east which house the working class, food markets and informal trade (referred to, by contrast, as 'The City of Slums').

This organisation relates to historic inequalities which have their roots in the colonial era and have been reinforced by public policies and private investment. For example, Altamirano (1995, p. 242) recognised in 1885 the unhealthy and deteriorated east of the city (La Merced), while in 1794 architect Ignacio de Castera of the New Spanish Baroque developed an urban plan designed to 'bring order' to the irregular and dirty neighbourhoods east of the city to correct and remove their wickedness. In these old visions, as in some contemporary ones, urban reform is at the service of social control.

Ribbeck (1991) recognises that the historical socio-spatial segregation of the historic centre of Mexico City is not static but rather very dynamic over time. For him, this urban bipartition is a struggle between two 'twin cities' that push from each side of the 'frontier' in attempts to conquer the neighbouring territory. Here the dispute is not between civilisation and barbarism but between the modern city and working class or traditional city. Thus, there have been periods when the pre-modern and working class city has pushed the frontier to conquer 'the City of Palaces' with their practices (slums, informal trade). There are other periods when the government and the elite, through public policy, return the border of the working class to its place, evicting hawkers and slums, revamping obsolete buildings in physical and economic terms; and sometimes trying out public policies to expand their frontier to conquer damaged working class territories and designate them for 'noble' and 'worthy' uses. These twin cities therefore coexist, mutually threatening each other with invasion. Currently public policies for urban heritage extend the border of the 'dignified', modern and orderly historical centre, towards the north (near Plaza Garibaldi) and east (La Merced) towards the historically working class and 'deteriorated' territories that are gradually tamed and conquered.

The model of 'recovery' of the historic centre

Since 1967 there have been eight generations of public policies for the 'recovery' of the historic centre of Mexico City. 'Recovery' and 'rescue' are concepts used prolifically by governments, advocates and restorers of the urban architectural heritage to describe the actions that they take to extract the built heritage from its state of deterioration, obsolescence and/or inappropriate use. These concepts (coupled with others such as rehabilitation, revitalisation, regeneration, remodelling) are used in a neutral and apolitical manner and therefore the semantic burden of these concepts usually goes unnoticed. There are few critical academics and opposition politicians who ask the question: For what and for whom is the 'rescue' of the historic centre for? Who decides what the economic, functional and physical obsolescence the programme of architectural heritage which necessitates its 'recovery'?

With the entrenchment of neoliberalism as an economic doctrine since the 1990s, public policies for the historic centre promote the active participation of the private sector. In this way, the 'recovery' of the urban landscape is part of the discourse for the entire population, dignifying the built heritage and strengthening national identity. But in practice private business, tourism, cultural consumption and the uses of urban heritage by higher-income groups are promoted. This vision stigmatises, criminalises and displaces certain social practices and people considered non grata or suspected of crime. This new urban order is based on five basic components:

1   Citizen 'participation' in the decision-making process: Based on a political agreement between the local and federal governments with the richest investor in the country, Carlos Slim, an Advisory Council for the Rescue of the Historic Centre was formed in 2002. The Mexican tycoon presides over this entity which does not consult anyone and which not a single resident or shopkeeper is part of. Carlos Slim, identified as a philanthropist who 'rescues' world heritage has, between 2002 and 2004, bought 63 properties which have become part of his real estate business (headquarters for his companies, housing and commercial rental space) (Delgadillo, 2016).

2   Physical improvement: Renovation of facades and public spaces, the replacement of infrastructure, street furniture, paving and pavements. In 2007 a new Metro-bus route was introduced, linking the historic centre with the international airport. Here, the owners of re-valued properties capture the increase in profits triggered by public investment which is not recovered by the state.

3   Increase in security: An increase in the number of police, surveillance video cameras and systems of citizen alarms[3] to ensure a suitable climate for real estate, commerce, service and visitors. In 2002 the former Mayor of New York, Rudolph Giuliani, and promoter of 'zero tolerance' advised on the public safety programme for the historic centre. Based on his recommendations, in 2004 the local left-leaning government issued the Civic Culture Law, which grants powers to local government to evict informal activities and those suspected of crime from the streets.

4   Relocation of street vendors: Since the 1990s there have been three generations of public programmes to relocate street vendors in shopping centres. The largest recent programme was undertaken in October 2007,

when 15,000 street vendors were relocated to 36 shopping centres[4] in the historic centre.

5   Expanding the boundaries of the recovery of the historic centre: In 2007 the programmes for the 'recovery' of the historic centre began to push the boundaries of the 'City of Palaces', to include two working class areas located in the north and east of the historic centre: the Plaza Garibaldi and its surroundings, converted into a theme park of mariachis and tequila (traditional music and drink); and the Old Merced, through the pedestrianisation of streets and the installation of modern shops intended for higher-income consumers. In addition, since 2013 the 'rescue' has been promoted in the market area of La Merced, to the east of the historic centre.

Public markets in the city's food supply system

For the supply of goods, Mexico City has 1,899 facilities employing 265,000 sellers. Public markets account for 17 per cent of these facilities and 27 per cent of their employees (see Table 2.1). In the following paragraphs we explain the different types of retail channels in the city.

Public markets

The current market regulations dating from 1951 define public markets as any place 'where there are a variety of merchants and consumers, whose supply and demand is concerned mainly with basic produce' (DOF, 1951). This definition is now restricted to covered structures specifically built for the purchase of goods with necessary spaces and facilities. The state, the owner of these structures, licenses and rents the market stalls. The local government department of the Ministry of Economic Development (SEDECO) oversees the functioning of the markets, while the administrative delegations[5] are responsible for their management and maintenance. Most of the public markets are located in the central areas of the city because since the 1970s no

Table 2.1. Traditional retail channels in Mexico City

Type	Quantity	Employees	Percentage of type	Percentage of employees
Public markets	329	71,985	17.32	27.22
Tianguis	1,303	171,820	68.62	64.96
'Wheeled' markets	52	1,483	2.74	0.56
'Concentrations'	215	19,204	11.32	7.26
Total	1,899	264,492	100.00	100.00

Source: Calculations based on data from SEDECO (2014a).

public markets have been built. The supply of the expanding periphery is achieved through other means.

Tianguis

The name of this type of market originates from the pre-Colombian indigenous Nahua language and refers to an open outdoor space intended for the purchase of goods and products. The Tianguis receive a permit to occupy public spaces such as squares, parks and streets. (For a discussion of the Tianguis in Mexico City, see Gómez, in this book.)

'Wheeled' markets

This market is an itinerant form of Tianguis where traders have the authorisation to occupy public space (squares and streets) once a week. These markets move their goods by car and occupy a different place in the same district every day, returning to the first location at the start of each week.

'Concentrations'

This type of market supplies the working class through an 'association of traders who trade in general products on public streets and lacks the most essential infrastructure for their proper functioning' (SEDECO, 2014a). This is informal trade that occupies the public realm. Informality refers to activities not regulated by the state which are not criminal, that is to say that they are lawful economic activities that use 'illegal' means.

Supermarkets

In Mexico City there are 332 supermarkets and 225 department stores (SEDECO, 2014a). Supermarkets, virtually unknown in Mexico until the 1970s, have grown in number since the enforcement of the free trade agreement between Canada, United States and Mexico in 1994. Contralínea (2010) characterises Mexico City as 'the City of Walmart', criticising the encouragement of the establishment of supermarkets by the local government, regardless of their effect on public markets. SEDECO (2015, p. 10) notes that in seven administrative districts there are more supermarkets than public markets.

Trends and problems in markets

The forms of consumption in the megacity of Mexico City have changed substantially in recent decades resulting from different economic, social and political factors (SEDECO, 2015). The neoliberal policies adopted by local and national governments gradually abandon traditional food distribution systems to the detriment of food sovereignty in favour of food imports

(Torres, 1999, 2003). In the city the growth in the establishment of national and international supermarkets has been favoured. These supermarkets vie for the customers of traditional public markets in a very unequal competition. For many governments the traditional channels of supply are inefficient, ineffective and a burden for the modernisation of supply and public markets are dilapidated and obsolete in physical, functional and economic terms and should therefore be renewed and modernised (Castillo, 2003). Based on a review of the problems of the various markets as reported in the press, we outline the following trends in the transformation of and disputes over public markets in Mexico City.

1   Unfair competition: Markets suffer differentially according to their geographical location and the pressure and competition from supermarket chains which offer the same products at cheaper prices, credit, credit card payments and car parking. They also deploy intense marketing campaigns.
2   Transformation from markets of basic goods, including food, to ready-made or takeaway food markets. Markets traditionally included an area for the preparation and sale of food for customers and traders themselves. However, for various reasons[6] this trend has now been exacerbated in certain markets, which have almost exclusively become food halls with accessible prices.
3   Declaration of obsolescence and decay: In the last decade there have been some diagnoses of present public markets as obsolete, deteriorated, insecure, vulnerable and / or functioning under irregular legal conditions. These features might not be far from reality but they are magnified according to specific economic or political interests. For example:

    • A local councillor said that in Mexico City 44 markets were at risk of collapse due to their age and that 68 per cent of the markets were dilapidated (Suárez, 2013).
    • A local councillor said that illegal products were sold in public markets and that one in three shops were abandoned (Cruz, 2009). This is supported by the head of the SEDECO who stated that 35 per cent of local public markets were abandoned or used as warehouses (Reforma, 2007a).

4   Proposals for the replacement of some public markets based on the apparent physical obsolescence of markets. For example:

    • The local councillor of the Benito Juárez district planned three urban projects which involved the conversion of public markets, considered obsolete, and their replacement with taller buildings (Quintero, 2007).
    • The National Chamber of Commerce offered to invest in public markets that were underused and had been colonised by street vendors. 260 public markets (generally one-storey structures) were offered the possibility of increasing more than three stories in order to relocate hawkers (Reforma, 2007b).

5   Selective transformation into gourmet markets: Recently there have been initiatives to transform some public markets in markets of gourmet food for tourists in emulation of the success of the market of La Boquería in Barcelona (Reforma, 2011) and the San Miguel market in Madrid (Suárez, 2014). (See also Salinas and Cordero's analysis of gourmet markets in Mexico City and Madrid, in this book.) Of course the presence of an existing market to create a gourmet market is not necessary: In 2014 in the area of Colonia Roma the Roma Market was created in an old industrial structure and in San Angel neighbourhood the Mercado Del Carmen was opened in a colonial house. Both 'markets' share the discourse and marketing of a 'new way of understanding the market' and as places to 'create unique experiences of social interaction'.

La Merced: A contested and problematic megaproject

La Merced is an urban area characterised by a mix of public markets, street vendors, urban congestion, public insecurity, physical deterioration, sex work, poverty and depopulation. Its name comes from a convent of the Mercedarian order that was almost totally destroyed in 1862 in order to build a new market to relocate ambulant traders and stallholders from a market that had itself been demolished. That nineteenth century market had spilled out into the surrounding streets through informal trading stalls, which created urban congestion and led to its demolition and relocation to the current market area of La Merced. In turn this collection of markets became saturated and a new Central de Abasto[7] (Central Market) was built. However, the markets of La Merced retained their metropolitan vocation, and new shopping centres and informal trade add to the strength of the commercial function of the area. SEDECO estimates that between 200,000 and 250,000 people come daily to La Merced to buy goods or work (SEDECO, 2014b, p. 10).

The markets of La Merced, built in 1957 (Cetto, 2011), are an icon of modern Mexican architecture and of the local government policy for the construction of markets at that time.[8] Originally there were five public markets (now eight) with 4,900 commercial premises (see Figure 2.1 and Table 2.2). Their construction, undertaken in eight months, employed an innovative technology for its time – a reinforced concrete structure with vaulted concrete roofs.

The unplanned markets are located in the 'underpass' tunnels that link the main and small market halls, The Banquetón market created on the west side of the main market hall to relocate those displaced by the construction of the Merced station for the Mexico City Metro, and the Ampudia market located in a four-storey building constructed in the 1940s.

Since the 1990s, 14 shopping centres have been built in the area of La Merced through a policy of relocation of informal trade that occupies the streets of the historic centre. Amongst the largest are the San Ciprian market

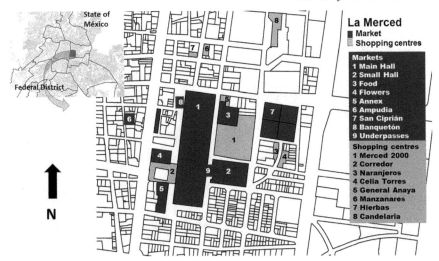

Figure 2.1 Location of La Merced Market
Source: Victor Delgadillo.

Table 2.2. Markets of La Merced

Mercado	Stalls	Area (m^2)	Speciality
Main hall	3,851	37,762	Fruit, vegetables, chillies, seeds, prickly pears, food, plastics
Small hall	555	8,174	Dairy, meat, fresh fish, seafood
Annex	186	8,036	Household and kitchen wares
Flowers	108	3,291	Artificial flowers, decoration, aquariums
Food	218	4,676	Food
Banquetón	401	2,807	Food, vegetables, clothes, chillies, seeds
Underpasses	73	2,637	Cleaning products, costumes, plastics, decoration
Ampudia	152	1,421	Confectionery

Source: The author.

and Merced 2000. The huge informal trading activity has surrounded the markets and expanded into surrounding streets and primary roads, obstructing pedestrian and vehicular traffic. The local authority has tolerated this activity, perhaps recognising the right to work and the lack of jobs, and has gradually regularised the occupation of public space. Thus the Santa Escuela street and the esplanade of the La Merced Metro station have practically disappeared. In the latter, the government has built a structure with a sheet metal roof to order and regulate the informal trade (Figure 2.2).

Figure 2.2 Roof of La Merced Market
Source: Víctor Delgadillo, 2015.

The issues and problems affecting the markets

The old, dilapidated, pre-Colombian and colonial neighbourhood of La Merced has become the periphery of Mexico City's historic centre. The commercial vocation that was assigned to the area in the nineteenth century was strengthened with the construction of the markets in 1957 and the shopping centres of the 1990s. These and the construction of other public buildings have been unable to reverse the stigma of insecurity and decline of the area which has historically tolerated sex work, street trading and the presence of a marginal population (homeless people, street children, drug addicts) who find the means for survival in this area. Below we list the most important struggles and issues currently affecting the complex of markets and various retail types in La Merced.

1   Formal traders versus informal trade: One of the potential problems for tenant stallholders and established formal traders is the presence of around 3,000 hawkers that surround the markets and spill over into the adjacent residential areas. Tenant stallholders accuse the local authorities of allowing the presence and expansion of street vendors that occupy the adjacent streets and markets, complaining that they constitute unfair competition which does not pay taxes, flood the streets with the same goods, occupy the loading and unloading areas and parking lots, and that they constitute a danger because they hinder the passage of emergency service vehicles. In an interview, Petra,[9] a tenant stallholder affected by the last fire in the main market hall, complained that the government tolerates the informal occupation of public space because it yields

economic and political gains (interview, 04/03/2015). Meanwhile the street vendors and their powerful corporate and pyramidal organisations defend the right to work in a city that does not create jobs. In addition, hundreds of young people without any training offer the only thing they have – their labour – to load goods as 'diableros'.[10]

2   Vehicular and pedestrian congestion: Due to the teeming occupation of streets, squares and pavements by informal traders and the large influx of buyers the market area is very congested by vehicles and people. In the markets there is a lack of loading and unloading areas and those that do exist have been occupied by informal trade. In an interview in April 2014 the administrator of the main market hall told us that many tenants have to unload their goods in distant streets and pay the 'diableros' to carry their produce to market.

3   Physical obsolescence: The collection of markets has lacked adequate and ongoing maintenance. Each market is in a different condition, but it is possible to generalise various situations. In general, the water, sanitary and electrical installations are in a deteriorated condition and the roofs leak. In addition two markets have suffered fires, the main hall in 1988 and 2013 and Ampudia in 2001 and 2014. According to the local government the fires are due to the informal sale of fireworks and irregular electrical wiring, while for the tenants the causes are the negligence of the authorities in charge of maintenance (Excelsior, 2013). The rehabilitation works to the main hall following a fire in February 2013 have not been completed (at the time of writing, April 2016); meanwhile tenants have been relocated 'provisionally' on the street.

4   Public insecurity: the insecurity of the area includes the poor lighting in the street and the markets, obstructions of the streets by informal trade and the refuse left in the streets. In a recent survey we carried out with 300 people from the La Merced neighbourhood,[11] 68.7 per cent of residents surveyed said that insecurity is either a serious or very serious problem, while 69.7 per cent said that street robberies are a serious or very serious problem.

The urban redevelopment megaproject

According to SEDECO the last fire in the main hall represented an 'opportunity' to modernise the area. Since 2013, the local government has promoted a 'comprehensive rescue' of the market area. With the same logic of 'participatory' decision-making, an Advisory Council for the Integral Rehabilitation of La Merced was set up. This council does not consult anyone and does not comprise any tenants of the markets, traders or residents of the neighbourhood. The presidency of the council has been given to a personality from 'civil society', a conservative journalist known for presenting the nightly news programme on the largest private television channel in the country.

In November 2013 a competition for conceptual ideas for the comprehensive 'rescue' of La Merced was held. The winning project uses politically correct and 'cutting-edge' language. It aims to increase the value of 'the public space and markets as facilitators of the commercial, cultural and social activities'[12] and as a 'trigger in the process of rebuilding the social fabric' to 'reconnect' adjacent neighbourhoods and improve the image, mobility, security, performance and liveability of the area. The project includes a new National Centre of Gastronomy, branch offices of banks, the creation of a huge new public square in the heart of La Merced (at the expense of the destruction of several buildings) and a network of pedestrian routes to 'increase the commercial potential of the area' (something that this area does not lack). The curious similarity of the concept of this project with the 'recovery' of the Raval district in Barcelona is surprising. In that city a new Rambla was created in a deteriorated neighbourhood as a mirror of the famous Rambla of the Ciutat Vella. In La Merced, the mirror is the Zócalo or Main Square of the city.

The project also aims to increase the building heights (from seven to twelve storeys) on the Fray Servando Teresa de Mier Street and designate the site as mixed residential use (a code that allows building plots to be used for virtually any activity). At the same time a different project plans to build 3,000 new homes in the area of San Pablo, in front of La Merced, through the replacement and repurposing of buildings.

The announcement of this megaproject has led to the various stakeholders (re)defining their vision of the future of the markets. There has been a general rejection of this project by informal traders and the tenants of the markets and shopping centres, particularly the Merced 2000 mall (which the winning project proposes to demolish), and especially at the beginning when a rumour spread that the markets would be destroyed to build a mall (Díaz, 2014).

In La Merced there are different social groups with varying degrees of organisation and with very different interests in the area: resident owners and tenants; established traders; stallholders in markets and shopping centres; ambulant trades; property and business owners; the destitute; indigenous people; the 'diableros' workers; religious organisations; sex workers; NGOs and so on. These actors have very diverse interests, which are often in opposition, competition or conflict, so it is no coincidence that while many of them converge in opposition against the recent La Merced Comprehensive Rescue Project, they do so in different ways.

The traders of the various markets publicly expressed their opposition to the megaproject during the three months that the project was exhibited by the winning architects of the competition during the so-called 'Forum for the Future' which took place at the City Museum in 2014. At one of these public forums, the stallholders of the flower market complained that the architects who had won the competition said that their market had no identity, leading them to ask, 'How are we to blame?' Traders have opposing views about the megaproject and the future of La Merced. For example:

- Petra, a trader affected by the most recent fire in the main market hall, supported the strategy of the megaproject to demolish the Merced 2000 shopping centre because of their bars, which offer a poor service, have no permission to be there and have nothing to do with the traditional markets (interview, 04/03/2015).

- Eusebia, a trader in the food preparation area of the main hall, in an interview commented that they are in favour of the megaproject provided that it first clears the streets, relocates the street vendors and ends the private contract for managing toilets in the main hall, which instead of being beneficial to all is a business for the administrator (interview, 04/03/2015).
- Three traders of the Flower Market, recognised the need to address the area but also for traders to participate and remain in the markets (interview, 04/03/2015).
- A trader in the under-used Merced 2000 shopping centre said that more should be done with less, that they should restore buildings instead of demolishing them, recover public spaces by relocating street vendors, police the markets and streets and illuminate them. For the Merced 2000 centre they proposed a radical transformation, with modern architecture and garish illuminated facades like those used in the USA or Europe in order to convert it into a key commercial reference so that Merced 2000 're-emerges as a modern building' (interview, July 2014).
- Silvestre, a trader that converted his stall which sold produce into a refreshment stand selling pre-prepared food, has on various occasions spoken of imminent displacement, of gentrification (he uses this concept), of the need for tourism and their desire to stay in La Merced, although to do so he has to change his line of business and 'leave behind the quesadillas to sell Spanish cheese and wines'. He built a second floor above his stall (somewhat unauthorised according to SEDECO officials) in order to open a cultural and educational space for the children of the traders. For him the 'comprehensive rescue' megaproject of La Merced threatens the displacement of the traditional traders and products by new traders of select products. Silvestre asked us to bring as many visitors to his stall as we can so that the authorities realise that La Merced, as it is, can attract many visitors without the need for rescue programs. 'Visitors can contribute to the defence of the market as it is now' (interview, 02/04/2014).

The megaproject is now on hold. This is due to three reasons: Firstly, strong opposition from tenants and ambulant vendors who, when it comes down to it, will carry out vigorous protests to oppose it; secondly, political circumstances – the elections of deputies and councillors in June 2015 and the election of the Constitutional Assembly in June 2016 which will ratify the Constitution of Mexico City as an autonomous entity; thirdly, economic projects which are much larger than that of La Merced, such as the new airport following the planned obsolescence of Mexico City's existing International Airport by the federal government.

In an interview, the Director General of public markets in the city commented to us that the competition was only for 'concept ideas of what to do with La Merced' and that the winning team will carry out a Masterplan for the full recovery of La Merced which will not necessarily include the architectural urban proposals presented in the winning project (interview, 14/04/2015). In this interview he told us that the Masterplan includes 108 short-, medium- and long-term projects. The three priority projects are: housing, which includes the improvement of the Candelaria de los Patos neighbourhood and a project in San Pablo; 'recovery' of public space in the semi-pedestrian Corregidora Street; and the improvement of commercial activity in the small market hall including the renovation and replacement of market stalls. Our interviewee commented that the modernisation of La Merced was 'urgent' and that those who oppose the project 'are informal vendors and irregular tenants who do not want things to change, who continue to take advantage of the situation and to sell things without authorisation'.

In March 2016 the only projects that are moving forward (without notification of the public) are the renovation of the small market hall which has caused disagreements with some of the tenants and the pedestrianisation of Corregidora Street, where negotiations with the leaders of the street vendors have begun. The group 'Left Hand Rotation', who worked intensively for three months with tenants and informal vendors of La Merced in 2015 demonstrate (from evidence collected from interviews) the widespread rejection of the megaproject and a constant complaint about the lack of information about the projects and public policies of local government (Left Hand Rotation, 2015).

Conclusions

The construction of a discourse of the deterioration and obsolescence of La Merced and the condemnation of certain social and economic practices (sex work, street vendors) are historic issues in Mexico City. These so-called indecent practices as well as the overflowing markets and the informal trade are recurrent themes in the history of the city, in the same way that street traders have often been relocated to beautify or 'rescue' public space and decongest the city centre.

The project for the 'comprehensive rescue' of La Merced neighbourhood and its markets is part of a dual policy for the preservation of urban heritage and the modernisation of the public retail network. The urban frontier of the 'recovered' historic centre, orderly, dignified and clean is extended to incorporate a territory historically considered abnormal and chaotic but one that contains a rich architectural and urban heritage and diverse economic activities. In the context of scarce colonial and nineteenth-century architectural heritage, the markets themselves become part of the heritage. The modernisation of markets promotes the incorporation of traditional sales transactions in formal banking mechanisms, the introduction of supermarket trolleys with the argument of promoting more shopping which appears directed towards the displacement of the other informal practices of the 'diableros' and the recognition of the local gastronomy which aims to incorporate the activities of markets in the industries of tourism and cultural consumption.

The language of the project for the 'comprehensive rescue' of La Merced, apparently neutral and depoliticised, legitimises the discourse of the 'natural' obsolescence of markets: the physical structures are damaged and constantly catch fire, the premises are obsolete and the traditional business model is not very competitive.

Many tenants, formal traders and informal vendors (with very different political inclinations and loyalties) consider the project of 'comprehensive rescue' as implying direct or indirect displacement through the destruction of buildings, the change of the use of the property and / or the type of products sold there. However, the authorities do not understand the social rejection of their 'good' intentions and resort to discrediting the opposition, arguing that their opponents have individual interests and are defending illicit activities. Many of our respondents recognise the need to confront the problems of La Merced, but demand their continued presence in the market and participation in decision-making.

The dispute over the markets of La Merced is the dispute over urban heritage and a model of society and of the city in its multiple economic, social, cultural and symbolic dimensions. The dispute over the public markets of La Merced is first and foremost the dispute over a type of market, one which is 'chaotic' for traders and working class consumers; and the other which is ordered, aseptic and hygienic for tourists and higher-income consumers.

Notes

1 Original in Spanish, translated by Neil Turnbull.
2 The terms 'rescue' and 'recovery' are in quotation marks to highlight the vocabulary used by the local government and other stakeholders. Its significance is explained later.
3 Citizens' alarms are located on posts in the street. In case of emergency citizens can activate the alarm button or a mechanism which alerts the police.
4 These shopping centres can occupy an entire multi-storey building or are installed on single-storey premises. In both cases, they are enclosed by either walls or fences.
5 Mexico City is divided into 16 administrative delegations.
6 For example: transportation costs and travel times (preventing many people from returning home to eat), the incorporation of women into waged work and the proliferation of convenience stores.
7 As part of a policy of the decongestion and conservation of the historic centre, the new Central de Abasto was established in Iztapalapa by the local authority in 1982.
8 These markets were built as part of public policy that between 1952 and 1966 built 88 markets and opposed and combatted street vending.
9 The names of the interviewees have been changed.
10 'Diablito' is the name given to a cart used to move goods. 'Diablero' is the name given to those who drive these carts.

34   *Víctor Delgadillo*

11 These surveys are part of a wider study of 10 areas of the city where 3,000 surveys conducted in August 2014 in 10 central areas by the Research Group *Habitat and Centrality*, funded by CONACYT.
12 The phrases in quotation marks come from the technical report of the winning project, published by SEDECO (2014b).

References

Altamirano, I.M. (1995 [1885]). *Paisajes y leyendas, tradiciones y costumbres de México*. México: Porrúa.

Castillo, H. (2003). Los mercados públicos de la Ciudad de México. Características, problemas y ¿soluciones? In Torres, G. (Ed.), *Políticas de abasto alimentario. Alternativas para el Distrito Federal*. México: UNAM – Juan Pablos, pp. 189–195.

Cetto, M. (2011 [1961]). *Arquitectura Moderna en México*. Edición Facsimilar. México: Museo de Arte Moderno – INBA – APASCO – CONACULTA.

Contralínea (2010). La Ciudad del WalMart. *Contralínea*, p. 204.

Cruz, A. (2009). Piratería en 317 mercados públicos. *El Sol de México*, 4 March 2009. Available from: http://www.oem.com.mx/esto/notas/n1070401.htm

Delgadillo, V. (2005). *Centros Históricos en América Latina: Riqueza patrimonial y pobreza social. La rehabilitación de vivienda en Buenos Aires, Ciudad de México y Quito*. PhD Dissertation. México: UNAM.

Delgadillo, V. (2014). *Estudio sobre el patrimonio urbano de La Merced*. Delegación Venustiano Carranza. México. Unpublished document.

Delgadillo, V. (2016). Selective modernization of Mexico City and its historic centre. Gentrification without displacement? *Urban Geography*, 37(8), 1154–1174.

Díaz, G.L. (2014). La Merced será de otros. *Proceso*, 27 March. Available from: http://www.proceso.com.mx/368208/la-merced-sera-para-otros-2

DOF (Diario Oficial de la Federación) (1951). Reglamento de Mercados para el Distrito Federal. México: Diario Oficial de la Federación. 1 June 1951.

Excelsior (2013). En 25 años incendios consumen La Merced. *Excelsior*, 12 December 2013. Available from: http://www.excelsior.com.mx/comunidad/2013/12/12/933474

Left Hand Rotation (2015). Permanecer en La Merced. Documentary film. Available from: https://permanecerenlamerced.wordpress.com/2016/02/24/permanecer-en-la-merced-documental/

Mongin, O. (2006). *La condición urbana, la ciudad a la hora de la mundialización*. Buenos Aires: Paidos.

Monnet, J. (1995). *Usos e imágenes del centro histórico de la ciudad de México*. México: CEMCA – DDF.

Quintero, J. (2007). Proyecto inmobiliario amenaza a mercados públicos en la Benito Juárez. *La Jornada*, 28 November 2007. Available from: http://www.jornada.unam.mx/2007/11/28/index.php?section=capital&article=043n1cap

Rapoport, A. (1972). *Vivienda y Cultura*. Barcelona: Gustavo Gili.

Reforma (2007a). Subutilizan locales en mercados públicos. *Reforma*, 18 December 2007. Available from: http://ciudadanosenred.com.mx/noticia/subutilizan-locales-en-mercados-publicos/

Reforma (2007b). Ofrece IP invertir en mercados públicos. *Reforma*, 19 December 2007. Available from: http://ciudadanosenred.com.mx/noticia/ofrece-ip-invertir-en-mercados-publicos/

Reforma (2011). Proyectan elevar rango a mercados. *Reforma*, 7 November 2011. Available from: http://ciudadanosenred.com.mx/noticia/proyectan-elevar-rango-a-mercados

Ribbeck, E. (1991). Mexico Stadt: City of hope, city of despair. In: Delgadillo, V. and Wemhöner, A. (Eds.), *Mexiko Stadt*, Stuttgart: Universität Stuttgart, pp. 17–25.

SEDECO – Secretaría de Desarrollo Económico del Gobierno del Distrito Federal (2014a). *Reporte Económico de la Ciudad de México*. Available from: http://reporteeconomico.sedecodf.gob.mx/

SEDECO – Secretaría de Desarrollo Económico del Gobierno del Distrito Federal (2014b). *100 Visiones sobre La Merced. Distrito Merced*. México: SEDECO – GDF.

SEDECO – Secretaría de Desarrollo Económico del Gobierno del Distrito Federal (2015). *Política de protección y fomento para los mercados públicos de la Ciudad de México (2013–2018)*. México: SEDECO.

Smith, N. (2012). *La nueva frontera urbana. Ciudad revanchista y gentrificación*. Madrid: Traficantes de sueños.

Suárez, M. (2013). En riesgo, 44 mercados públicos por su antigüedad y daño estructural. *La Jornada*, 18 January 2013, p. 40.

Suárez, G. (2014). La otra cara del mercado. *El Universal*, 23 March 2014. Available from: http://archivo.eluniversal.com.mx/ciudad-metropoli/2014/impreso/la-nueva-cara-del-mercado-122320.html

Torres, F. (1999). *Alimentación y abasto en la Ciudad de México y su zona metropolitana*. México: GDF – UNAM.

Torres, G. (Coordinador). (2003). *Políticas de abasto alimentario. Alternativas para el Distrito Federal*. México: UNAM – Casa Juan Pablos.

3 Learning from La Vega Central

Challenges to the survival of a publicly used (private) marketplace[1]

Elke Schlack, Neil Turnbull and María Jesús Arce Sánchez

Introduction

This chapter focuses on the pressures of gentrification which threaten La Vega Central, a market privately owned by its stallholders in Santiago de Chile. Here the threat of gentrification is a process which benefits the affluent users of the market at the expense of the poor, encouraging a significant change in the social profile of its users. This in turn challenges the public characteristic of the market as understood as a universal space for all (Habermas, 1990). Given that the market already represents the phenomenon of privatisation of what was previously a public asset, any modification of its public role is linked to the debate about exclusion and the public realm (Weber, 1922). Public space becomes a lens by which the issues of access to the market and the right to the city debate as discussed by Marcuse (2009) can be explored.

When spaces of commercial exchange are privatised there is a challenge to the continuity of their contribution to the public space of the city. Public space has long been associated with markets and they are in this sense a spatial archetype of Western urban culture and represent in the field of urban sociology the construction of the public sphere (Weber, 1922). Using this idea, we reflect on the traditional indoor market of La Vega Central, placing emphasis on the way in which it produces public space. Significantly, La Vega is administered by an association of stallholder tenants who engender a character of publicness which is more elusive in the other private retail spaces of supermarkets and shopping malls in Santiago de Chile. In addition, the surroundings of this market constitute a shelter for vulnerable groups because of the proximity and availability of accommodation, work opportunities and low food prices.

As landowners, the condition of la Vega Central's traders stands in stark contrast to the contestation of market traders in parts of the Global South who struggle for the right to be present and make a living on the public streets of the inner city (see Skinner, 2008; Bromley and Mackie, 2009, Brown et al., 2010). For this same reason the situation is also different from the pressures of trader displacement through increasing rents in indoor markets, seen in parts of Europe (González and Waley, 2013). The risk of gentrification in the case of La Vega Central is the expulsion of street vendors, informal workers, beggars and the more vulnerable consumers whose presence today defines the publicness of the market.

La Vega Central is an indoor market with 1,041 individual stalls and 889 formally-registered traders (Servicio de Impuestos Internos, 2014), which support a population of around 2,100 stall traders and 400 workers employed in the wholesale areas. These 'formal' workers are joined in the market by around 120 hawkers and porters who sell empanadas and *humitas* (a boiled corn snack), dispense hot drinks from converted supermarket trolleys or make a living by transporting goods for a small fee. The market serves around 20,000 customers daily.

The aim of this chapter is to outline the current public role of La Vega, which whilst having survived processes of privatisation, nonetheless remains contested as those on lower incomes are threatened with displacement through processes of gentrification. We argue that the current accessibility of the market is being challenged through processes of urban renewal and the eradication of the affordable landscape surrounding the market. In addition to this threat the attractions of the market are being transformed and re-orientated towards a public with greater purchasing power as it enters into the logic of tourism.

Through analysis of surveys and interviews with the users, workers and the stallholders of La Vega and a review of local newspaper reports we will examine two main issues:

- How the traders are securing ownership of the land, contesting their potential displacement from the city centre and manage its public character through offering accessibility and activity for a diverse public.
- The emergence of three main threats: the risk of the commodification of authenticity (Zukin, 2010); the danger of tourism; and the impact of regulatory plans which promote inner city urban renewal, tourism and displacement of the existing lower-income population.

To conclude we discuss what implications this could have for urban policy.

The current ownership model of La Vega Central

When considered from a historical perspective, the 'private' status of trade has always existed. Studies of medieval squares show that markets commonly took place on private land and that these 'private' medieval markets were intended for public use while being operated by a private owner, or were the subject to feudal control (Berding et al., 2010; Guárdia and Oyón, 2015). In these cases the right of public access concurred with the wishes of the owner who defined that the space would be for public use (Selle, 2003; Siebel, 2000, 2007; Wehrheim, 2007, 2015).

In the case of La Vega, the public condition of the market has evolved over time. At the end of the nineteenth century, new structures to house the market were built on a large plot of land donated by a private owner and the 'Association of Businesses of La Vega Central' was created in 1895 as a private company in order to manage it. The market was in close proximity to the marginalised poor who lived in the adjacent boarding houses and rented rooms and who were subject to renewal strategies and campaigns to tackle the unsanitary conditions (Castillo, 2014). By 1930 the market was under the ownership of the municipality and became the most important wholesale and retail supplier of agricultural produce in the city (Bastías et al., 2011). This dominant position in the supply chain was challenged in the 1960s with the construction of larger distribution facilities as part of a modernisation programme for food distribution (I.M. de Santiago, 1961) and the introduction of supermarkets (ODEPA, 2002). These changes brought about economic decline by removing part of the market's wholesale business and increasing competition to its retail trade. By the 1980s national economic policies to reduce municipal costs led to the market being put up for sale, placing its future in doubt. At the time there was growing interest in developing urban policies to promote the repopulation of the inner city through urban renewal and there was a risk that new owners might find greater profit in the sale of land rather than in the running of an 'obsolete' market. This threatened the displacement of the traders from their place of work and the removal of a centre of supply for the inhabitants of the inner city. However, traders confronted this privatisation and the uncertainties that it represented by organising themselves into a collective of approximately one thousand of the existing tenants, each obtaining the right of domain to a percentage of the total area (Bastías et al., 2011). This is the basis for the current ownership model of La Vega Central, where an organisation of traders administers their own management of the market as proprietors of the land.

Conflict as a privately-owned market

While ownership does give the stallholders power over their own land, it does not automatically guarantee the continued existence of the market itself. This became apparent in 2001 when the local municipality imposed restrictive night-time hours of access for the delivery of goods and prohibited the use of forklift trucks in the vicinity of the market impacting upon its functional operation. This led to days of protests by the traders who saw this as a direct threat to La Vega. A declaration in the press by the national sub-secretary for transport, claiming to be supported by the Ministry of Housing and Planning the Ministry of National Assets and the local mayor, stated that the wholesale of produce should not take place in the city centre, that the larger distribution facilities on the periphery of the city should do this job and that government efforts were necessary to reduce vehicle congestion and the levels of pollution in greater Santiago (El Mercurio, 2002). The traders responded by stating that they wanted to maintain the market where it was, pointing out that it provided employment for over 20,000 people. An agreement was finally reached but the traders were left under no illusion that they were not welcome in the city centre.

Securing political support for La Vega

More recently there has been a change in the relationship between the market and the political establishment, as La Vega has become a populist stage for national politicians. A review of newspaper reports reveals that between 2010 and 2013 at least 18 visits to La Vega were undertaken by government ministers who were involved in delivering either explicitly political messages or participating in softer propaganda such as watching Chile's football team play in the World Cup and attending events for International Women's Day (see El Mercurio, 2010; 2012a; 2012b; 2013a). These associations suggest that the traders have been actively building relationships with the political class in a bid to win over their support to protect their position in the city. One emblematic visit took place in October 2013 when the then President of Chile, having given a speech on the direction of his final months in office, handed over to the traders land titles to a small part of the market that they had appropriated but which was still owned by the State (El Mercurio, 2013c).

It is useful here to reflect on the trajectory of the status of the traders themselves by using Marcuse's (2009) categorisation of power in relation to the right to the city debate. In his analysis of Lefebvre's right to the city, Marcuse (2009) underlines the importance of understanding the distinct opportunities different groups have to exercise their rights. In broad terms these can be outlined by three situations: (i) A demand for rights by those who in economic terms can be described as the excluded or parts of the working class who are deprived of material and legal rights and may also include those directly oppressed through cultural exclusion; (ii) an aspiration for greater rights by the alienated, in order to be able to lead a satisfying life, who may include small business owners and occasionally some of the gentry; (iii) the practice of already-established existing rights by the gentry, capitalists, establishment intelligentsia and the politically powerful (Marcuse, 2009). The change of the status of the traders would therefore see them move from a position where they demanded their rights as the working class tenants of the publicly-owned marketplace up until the 1980s to a position where they aspired for greater rights as small business owners and landowners of the market. In this sense the traders of La Vega have consolidated some basic rights to work in the city.

The inclusion of vulnerable groups in Chilean markets

At the same time as the traders have obtained certain rights, they have also taken on the responsibility and power to manage a space which has a long trajectory as an important component of the public life of the city and especially one which addresses the needs of the most vulnerable. In Chile one of the essential characteristics of the historic marketplace has been the active role of the poor in the production of this public place based on their own working class culture (Salazar, 2003). In the historic Chilean market, the practice of trade has coexisted with an 'underworld' of popular culture developed in taverns, gaming houses, including folk dancing and drinking (Bastías et al., 2011). The market is therefore not simply a place to which the working class have access, but becomes an important place for the development of public life suited to the needs of those who would otherwise be excluded.

Public accessibility and public use of La Vega

Universal accessibility is a condition that defines a truly public space to which everyone has the right of access (Habermas, 1990). This discourse conforms to the ideal of public space which can promote encounters between strangers and initiate meaningful associations (Fincher and Iveson, 2008; Valentine, 2008). This refers to the possibility of the enactment of free communication between people whilst at the same time this activity is dominated by an individual or group which exercises their control over the space (Habermas, 1990). Wehrheim (2015) states that it is essential to understand how this control over space determines the ways in which it is used, while also recognising the role of its spatial characteristics.

Any study of the public character of a market needs to examine the accessibility and management of the space in terms of who the intended 'target' customers are and what activities and uses are employed to attract them. In addition, analysis of who is actually in the market, why they are there and how they access the market provides information on the characteristics which help describe the public function of the space. This dimension of 'publicness' of La Vega and other markets is part of the analytical sphere already advanced in the introduction of this book which sees markets as spaces for sociability and where actors can develop meaningful social relationships, though not without tension and conflict.

The traders' 'target audience' and their openness to vulnerable workers

Responsibility for the management of the market rests with the traders and it is they who play a key role in determining the profile of the 'public' which uses it. Interviews with the stallholders reveal their intention to welcome a wide variety of users from the *pituquitas* ('posh' women from wealthy families) to the lower-income consumers. According to the stallholders, the diverse range of customers are attracted by the quality, variety and low cost of the produce, as much as by the friendly and individualised service. Significantly traders also mention the participation of the homeless and destitute in the public life of the market through acts of solidarity. This aspect could be described as a practice that provides an important safety net for the poor. As one stallholder comments:

[P]eople who are in bad way, they have La Vega, it has vegetables, fruit, everything ... some [people] come for the fruit, I would say every day. A piece of fruit that is a little damaged and [the stallholders] leave it to the side and people come to look through the boxes later.

(Interview with stallholder, 2014)

As stallholders mention, the market is an important place not only for the most marginalised who require food for their survival, but also the alienated in society, as a place of work:

Look, the market is very supportive, if someone arrives here who has done something wrong, made a mistake say, but wants to turn things around, it is possible here in La Vega. No one will ask them about their past, if you want to work, if you want to make a living, come on then, let's go.

(Interview with stallholder 2014)

The background of stallholders is relevant in this respect; Traditionally stallholders were economic migrants to the city who came to make a living through the sale of their produce, and these origins along with the traders' recent experiences in contesting privatisation means that they may be more likely to empathise with the working class.

The actual public in La Vega

In order to examine the actual public present in the market, why they are there and how they access it, a detailed survey of 200 users was undertaken. Data collected reveal that there is a wide range of users with some living nearby who arrive on foot while others drive by car from more distant parts of the city. Interviewees had very diverse educational backgrounds, ranging from people with little formal education, who tend to be immigrants, homeless, beggars, to those with a university education.

The presence of a diverse public in La Vega is supported by its spatial characteristics. The building itself is permeable and connected to the surrounding streets, leading to an increased footfall resulting from the movement of pedestrians taking short cuts through the market, and, in doing so contributing to its public nature. The majority of the surrounding inhabitants are generally on lower incomes and their close proximity to the market is important as they can take advantage of the possibility of purchasing goods at low prices and can avoid paying transport costs, since they can walk there. The spatial proximity of this group is due to the opportunities that they find in the surrounding area. The neighbourhood provides a variety of spaces which are accessible to those on lower wages, such as low-cost accommodation in the adjacent rooming houses of the *cités*[2] and numerous affordable spaces for leisure, entertainment and social interaction (Arce, 2015).

While at the neighbourhood scale there is support for those on lower incomes, at the city scale, it is those on higher incomes that are catered for. The market is accessible to those in other parts of the city through the use of private transport. The market's proximity to major trunk roads allow shoppers access by car and the recent construction of an urban highway connects the market with the more affluent inhabitants in Santiago's eastern neighbourhoods.

Activities and uses that attract customers and workers to La Vega

The principal reason users give for visiting La Vega Central is to buy goods. Customers are attracted by the low prices and the variety of the products on offer, such as fresh fruit and vegetables, cheese, dried fruit and nuts, meat, frozen foods and hard-to-find produce from other parts of Latin America. Furthermore, the produce is seen as contributing to a sense of quality and fairness that the market represents, which is not limited to the products it sells but also as a place of work, as expressed by one customer:

> People that seek quality food have some kind of idea that makes them refuse to go to a supermarket ... that is one of the big reasons why I will not, or rather why I will go to La Vega and not a supermarket. I don't like the concept of the supermarket, of big business, of financial gain and low salaries for the workers.
>
> (Interview with customer, 2014)

Another aspect which attracts people to the market is the atmosphere created by the traders. The traders' congenial relationships with the customers turn shopping into a friendly and enjoyable task, where going to La Vega is seen as a leisure pursuit. This attraction was understood by one customer as the intangible heritage of the market:

> This is a place, I think, that should be preserved and preserved precisely because it has a character ... I would say that shows in the relationship between users and vendors that you can't find elsewhere, and I think this is heritage, or rather I think those things should be preserved.
>
> (Interview with customer, 2014)

The existence of less privileged actors in the market is evident and the appeal of the market for these groups underlines an important relationship between the market and its neighbourhood and in part confirms the stated intentions of the market traders in practising solidarity with the excluded:

> We [the homeless] move around more at night, by day we are here very close to La Vega because here they give us food, because as people here say 'After God, La Vega'. La Vega is very good here, they give us food all

day [breakfast, lunch, dinner and snacks of fruit], be it during the week or at the weekend.

(Interview with homeless person, 2014)

The market is also a place where street vendors benefit from living close by; 'I am here every day with my juice stand and the clientele is good. Moreover I come from close by [From a street seven blocks from the market], so I don't spend anything to get here not even on a bus ride' (Interview with Peruvian street vendor, 2014; see Figure 3.1).

Finally, the convenience of the market is revealed by one local resident:

> I come every day to La Vega to shop as I live close by. If I was missing some ingredient for *escabeche* (typical Peruvian dish of chicken or fish) I'd pop out and would go to La Vega to buy it. It's great to live so close, we are like neighbours, some days I go twice a day or more, today I have been and I'm sure to go back later with my son when I collect him from school, we often both pass through the market without buying a thing, it's like going on an outing.

(Interview with Peruvian shopper, 2014)

Figure 3.1 Street vendors in La Vega Market benefit from close proximity to the market
Source: Neil Turnbull, 2015.

The market is accessible to a wide spectrum of the poor and the vulnerable. By returning to Marcuse's (2009) previously outlined categorisation of different actors in relation to the right to the city debate it can be said that much of the public present in La Vega share similarities with the group who Marcuse says is engaged in a demand for their rights. Within this group we can identify those who are excluded economically and culturally: the informal workers, porters, ambulant traders and lower-income customers of La Vega are indicative of the economically-excluded in class terms, surviving at the margins of society; the immigrants from other Latin American countries who work in the market are indicative of those who may be oppressed in cultural terms along lines of race or ethnicity.

The potential threats to the public of La Vega

The market's current conditions allow for access by a diverse population and significantly a group who are often oppressed and excluded. However, this 'public' is vulnerable to emerging processes which if fully developed will turn the market into a place for the affluent at the exclusion of the poor. These processes are influenced by both the traders of La Vega as managers of its space and the state through the actions of municipalities who manage the surrounding neighbourhood. The traders' operation of the market along traditional lines offering a friendly personalised service may have the unintended consequence of attracting tourists and is in danger of becoming commodified as a mark of 'authenticity'. This could re-orientate the market towards a public with greater purchasing power having the potential to transform the retail offer and exclude the poor. In addition the poor and the working class who live next to the market are vulnerable to displacement through state-led urban renewal and the promotion of tourism which would lead to the eradication of the supporting landscape which surrounds La Vega.

Commodification of authenticity and the emergence of tourism

Zukin (2010) writes about how authenticity has become a tool in the gentrification of some neighbourhoods in New York City. The appeal of these initially down-at-heel environments is based on the opportunity to experience the 'authentic' city through its old brick buildings, cobblestone streets and lively crowds while food stores offer 'authentic' freshly-picked produce (Zukin, 2010). Due to these characteristics, these neighbourhoods become renowned in a process similar to Bourdieu's (1984) theory of 'distinction' where preferences for consumption are dominated by the ruling classes. The authenticity of a neighbourhood is at first recognised by the culturally-aware who may not be economically wealthy but are typically able to pay higher prices. Their presence brings change in both the land market, with higher rents, and in the retail landscape, with higher prices and more exclusive products which cater to their needs (Zukin, 2010). These beginnings have many parallels to classic residential gentrification where urban pioneers invest in neighbourhoods, leading to a snowball effect, in turn leading to the displacement of residents (Lees et al., 2008). Zukin et al. (2009) have outlined what this can mean for the retail landscape by documenting the process of the appearance of high-end products suited to more affluent customers in new 'boutiques' which contrast with the disappearance of older stores which cater to a poorer, more traditional and less mobile customer.

Interviews with the customers of La Vega Central reveal that its authenticity is its foremost attraction. This authenticity is produced by the workers who provide a personalised service, offer quality produce at affordable prices and animate the space with raised voices, jokes and friendly repartee between themselves and with the customers, all culminating in a convivial atmosphere evocative of the traditional marketplace.

Currently the emergence of more exclusive products which cater to the needs of the more affluent does not yet appear to have occurred. The recent appearance of exotic produce which originates in other parts of South America is arguably a new phenomenon which has a mark of distinction and the 'boom' in ethnic restaurants (El Mercurio, 2013b) points towards the successful cultural appropriation of these products. However, this does not necessarily translate into the appearance of more affluent consumers in the market. Interviews with some of the stallholders who sell these products reveal that their main customers are high-end restaurants. So while their produce is consumed by Santiago's elite who flock to the city's new high-end ethnic restaurants or enjoy typical dishes at home, the produce finds its way to these tables through the working class kitchen staff or the immigrant live-in maids who serve the rich, and it is these actors who are present in La Vega. At the same time these goods are also consumed by the immigrants themselves as an important part of their identity.

Whether or not the authenticity of the market is now attracting more middle class consumers is tricky to quantify, but there are some signs that small scale reinvestment in the market is taking place. A survey of the market undertaken in 2004 indicates that 10 per cent of the stalls were abandoned (Ducci, 2009), whereas field observations reveal that in 2014 only 1 per cent of stalls were disused or abandoned. A newspaper report from 2013 contributes to this narrative of renovation, with the administrator of the market commenting on the recent rise in the number of upper class customers. 'Before, the middle class came to La Vega, but they didn't amount to more than 2000 per week. Now around 15,000 people come from the A, B and C1 socio-economic groups. It's a big cultural change' (*La Tercera*, 2013). The report puts this increase down in part to customers choosing the market for its sense of heritage, emphasising the importance of the cultural authenticity of the market. (See Figure 3.2.)

While the increase in middle class footfall is a strong possibility, the rise in media focus the authenticity of the market is clear. In a review of newspaper articles in a national Chilean newspaper during the period 2000–2014, more than 18 different reports, all in the last seven years, refer to the cultural value

Figure 3.2 Foods from other parts of South America are on sale to both high-end
         restaurants and migrants
Source: Neil Turnbull, 2011.

of the market, including: the promotion of the market as a place that is
recognised by artists for its marginality; its adulation by musicians as one of
the cradles of popular Chilean music and numerous televised programmes
from late night chat shows, daytime soap operas and charity telethons (see El
Mercurio, 2014a; 2014b). Many of these aspects have parallels to the staged
events undertaken by shopping malls (Crawford, 1992), however here they
combine with the everyday attractions of the market to create a powerful
attraction based on tradition designed to draw in local customers.

Recently, recognition of the traditional ambience of the market has become
international as web-based cultural pundits recently voted La Vega the fourth
best market in the world, due not only to the opportunity to taste local foods
but also because it is a 'true reflection of the culture of the country' (The
Daily Meal website, n.d.). La Vega has been catapulted onto the international
tourist scene, alongside Barcelona's La Boquería, London's Borough Market
and Seoul's Noryangjin Fish Market, who take the first three positions in the
list (see also Salinas and Cordero's examination of the role of international
gourmet markets in gentrification in Chapter 6 in this book). This development relates to the local character of the place being captured by the media
which puts the market in danger of becoming reduced to a 'cultural destination'
(Zukin, 2010).

In other contexts, these processes which attract the middle class to the market might combine with a rise in the land value of the market, putting pressure on the traders themselves and undermining the public accessibility of the space. However, here in the context of La Vega the threat comes from the potential for change in the product offer as the traders are the owners themselves. As more middle class people are attracted to the market the stallholders could take advantage of this group through raising prices. They may also modify their offer to cater for the needs of the passing tourist, as can be seen in the case of the Mercado Central, Santiago's traditional fish market which sells expensive lunches and snack sized punnets of fresh fruit (Pádua Carrieri et al., 2012). This may provoke processes of displacement similar to part of Marcuse's (1985) definition, but in this case related to product offer rather than land value, with 'direct displacement' as potential result as lower-income customers are priced out and the 'pressure of displacement' is invoked, as the goods offered to tourists may not suit the needs of the regular customers.

State policy for urban renewal and tourism

While the traders are responsible for the accessibility of the marketplace itself, wider processes are at play which impact upon the surrounding neighbour-hood. Three municipalities (Recoleta, Independencia and Santiago Centro) are the custodians of the land around the market and the related warehouses and wholesale trade. These municipalities have different approaches to the market; however, none of their future projections considers strategies to address the protection of the Vega Central as an inclusive public space (Arce, 2015) to which we refer in this chapter.

The current urban planning ordinances of the municipalities are ambivalent and undefined in terms of land use regulation and of the future vision they have for the surrounding area including for the warehouses, related wholesale trade and the residential sector. On the one hand, there are some regulations that recognise existing land uses and try to coordinate the complex logistics involving market flows and the location of warehouses and trade. On the other hand, some regulations propose a change of land use from one which currently combines residential, warehouses and commercial use to one which is predominately residential. This ambiguity in the regulation of activities that are associated with the market leaves the traders and all those who undertake supporting activities in its vicinity exposed to a high degree of instability (Interviews with local planning consultants, 2014).

Vital components in the current activity of the market are the existing warehouses, and although they can generate negative external impacts on the functioning of the neighbourhood, it is necessary to acknowledge their role and look for ways to mitigate the issues. It is essential to find a balance between retail and wholesale activities, commercial infrastructure and housing that currently exists within the area where the consumers and the workers of La Vega reside (Arce, 2015). In contrast, policy outlined in the municipal regulatory plans, promotes the intensification of the land use through densification, in particular in the area where the most vulnerable housing has been historically located. In these areas high-density housing developments are permitted which allows for buildings 30 metres tall that have high floor area ratios between 1.5 and 4.0 (I.M. de Independencia, 2014a) and between 2.4 and 4.0 (I.M. de Recoleta, 2006).

These regulations combined with the state-sponsored subsidy for urban renewal create a lucrative rent gap for private investors and promotes the construction of residential buildings and are responsible for gentrification in the centre of Santiago (see Borsdorf and Hidalgo, 2013; López-Morales, 2011). In the case of the neighbourhood surrounding La Vega, an important determining factor in this process is the ownership model of these buildings where properties are held by a few owners who rent them out to lower-income tenants, leaving them vulnerable to displacement by landlords who wish to capitalise on the value of their land. These conditions have turned the area around the Central Vega into a coveted area for real estate development. This condition, alongside the absence of a policy to stop the risks of expulsion of the most vulnerable people, threatens the neighbourhood around La Vega, endangering both its working population and lower-income customers.

The municipalities that regulate the neighbourhood of La Vega demonstrate a certain ambivalence in its management. On the one hand they claim to be unable to make progress in addressing the issues associated with the market, such as traffic congestion and the attraction of informal trade to the surrounding streets, which they find difficult to control as they would like (Interviews with local planning consultants, 2014). On the other hand, they support initiatives to promote tourism in the area and the regeneration of the city centre for the middle class (Gobierno Regional de Santiago, 2014; IEUT and CCHC, 2016).

Reconciling the divergent interests of different actors and people present in the area (traders, residents, immigrants, Chileans and the transient population) is not at the heart of the discussion for the future of the area and the only attempts at coordination relate to the resolution of conflicts that arise from the activity of the market in relation to the use of the surrounding streets and pavements (I.M. Independencia, 2014b).

The three municipalities propose tourism as a future for the sector. There are even talks about the potential of this area, due to its centrality, to become a place of high-end products (Gobierno Regional de Santiago, 2014; Pujol, 2016). In this sense, the market's historic heritage and multi-cultural characteristics are seen as tools to strengthen the tourism sector through a number of projects, among which is the 'Esplanade of Markets'. This project corresponds to an International Competition from 2015, whose criteria asked for the design of an urban-landscaping project with emphasis on the integration of the two banks of the Mapocho River through a tourist hub (Arce, 2015).

In the various remodelling projects that the three municipalities plan, none addresses the concern about the inclusion and protection of vulnerable people and workers. Issues such as the improvement of pavements, parking for customers, the formalisation of tourist routes, the remodelling of heritage buildings and facades are promoted, as are the remodelling of public spaces for civic character (I.M. de Santiago, 2016).

Conclusions

The focus on the gentrification of public space allows us to address one of the central characteristics of indoor markets, that is to say their role in the public life of the city as a space of inclusion. In the case of La Vega, its specific ownership model is an example of the successful defence of traders' rights to contest displacement from their place of work and has allowed them to continue the tradition of the market in Santiago's city centre. By becoming landowners the organisation of traders of La Vega have not only to some extent safeguarded their rights to work in the city but also those of other workers associated with the operation of the market and other vulnerable users of the market such as the lower-income customers and those who benefit from the traders' acts of solidarity.

Nonetheless, this chapter reveals that the public nature of the market remains contested, even if the right of domain by the traders has been secured, with the rights of some members of the public to access the market potentially eroded. In practice the 'public' presence of the different communities of the affluent and the poor is nuanced and there are underlying potential conflicts between the two groups.

The findings from our research suggest that there is the risk of commodification of the authenticity of the market which has the potential to economically and culturally displace the current working class consumers. There is the risk that the absence of policies which safeguard the presence of the informal workers and immigrants in the surrounding neighbourhood will lead to their physical displacement through urban renewal. These threats challenge the continued role of the market as a public space and, in particular, call for a different response from the actors who exert an influence over its operation, directly in the case of the traders and indirectly in the case of the state.

Various questions are raised: To what extent can the traders offer an inclusive environment in the face of state policy to gentrify the area through tourism? Would the traders be able or willing to use their political contacts to advocate for the rights of the more vulnerable users of La Vega as they did for their own survival? How might the different local municipalities encourage development of the area and recognise the needs of the poor and the vulnerable, ensuring their participation in the city centre? And how might urban policy recognise these needs in the context of the development of markets?

Knowledge of how the most vulnerable shop and work at La Vega demands a deeper reflection on how the state and local government's public policy should address the area. Although they may see tourism as an interesting niche to exploit and the importance of promoting urban renewal of this sector of Santiago, we recognise that in the ecology of La Vega and its surroundings there are several groups who could be at risk due to these actions. If these actors are in danger so too is an important public space in Santiago which attracts tourists and people from around the city precisely because of these qualities. The challenge will be to develop the area on the understanding that the market's activity includes the surrounding whole-sale commercial activities, the warehouses, and the spaces where the lower-income traders and customers use of recreation of its workers in buildings and in the public space. The challenge will be to develop a new urban environment where the poor and the vulnerable are housed, have access to services and places for leisure. These might be imagined as new typologies of mixed-use buildings which house storage and offer cheap accommodation (Arce, 2015), accessible social housing in the city centre and affordable public use buildings for sport, community activities and education.

Finally, this is not merely a call for the preservation of the universal public space of the traditional market in the city but also a demand for the rights of the poor to be recognised by those who are responsible for the management of the city.

Notes

1 This chapter is part of the FONDECYT Regular No1120823 Chilean state-funded research project, 'Lo público y lo privado en la producción de espacios públicos vitales. Coproducción de espacios urbanos de apoyo en Santiago de Chile' (Private and Public Production of vital Public Space. The Coproduction of Supportive Urban Spaces in Santiago de Chile), Santiago, Chile.
2 *Cités* are a typology of collective housing from the early twentieth century.

References

Arce, M.J. (2015). *Aprendiendo de la Vega. Vitalidad como detonante proyectual.* Thesis presented at the Magister on Urban Projects (Magister de Proyecto Urbano) at the School of Architecture; Pontificia Universidad Católica de Chile.

Bastías de la Maza, C., Hayden Gallo, C. and Ibáñez Carvajal, D. (2011). *Mujeres de la Vega: Género, Memoria y Trabajo en la Vega Central de Santiago.* Santiago: Fondart.

Berding, U., Havemann, A., Pegels, J. and Perenthaler, B. (Eds.) (2010). *Stadträume in Spannungsfeldern. Plätze, parks und promenaden im Schnittbereich öffentlicher und privater Aktivitäten.* Detmold: Rohn.

Borsdorf, A. and Hidalgo, R. (2013). Revitalization and tugurization in the historical centre of Santiago de Chile. *CITIES*, 31, 96–104.

Bourdieu, P. (1984). *Distinction: A Social Critique of the Judgement of Taste.* London: Routledge & Kegan Paul

Bromley, R.D.F. and Mackie, P.K. (2009). Displacement and the new spaces for informal trade in the Latin American city centre. *Urban Studies*, 46(7), 1485–1506.

Brown, A.M.B., Lyons, M. and Dankoco, I. (2010). Street traders and the emerging spaces for urban voice and citizenship in African cities. *Urban Studies*, 47(3), 666–683.

Castillo, S. (2014). *El río Mapocho y sus riberas: espacio público e intervención urbana en Santiago de Chile (1885–1918).* Santiago: Ediciones Universidad Alberto Hurtado.

Crawford, M. (1992). The world in a shopping mall. In Sorkin, M. (Ed.), *Variations on a Theme Park: The New American City and the End of Public Space*. New York: Hill and Wang, pp. 3–30.

Ducci, M. (2009). *La Vega Central-El Mercado de Santiago, problemática y propuesta*. Santiago: Instituto de Estudios Urbanos y Territoriales, Universidad Católica de Chile

El Mercurio (2002). Estipulado horario definitivo de carga y descarga en La Vega Central. *El Mercurio*, 15 February 2002.

El Mercurio (2010). En La Vega y en la cárcel autoridades se darán cita para ver el debut de Chile en Sudáfrica. *El Mercurio*, 15 June 2010.

El Mercurio (2012a). Ministro (S) del Trabajo destaca que casi 640 mil personas encontraron empleo en este Gobierno. *El Mercurio*, 31 January 2012.

El Mercurio (2012b). Larraín destaca creación de empleos pese a situación internacional complicada. *El Mercurio*, 28 February 2012.

El Mercurio (2013a). Candidatos presidenciales salen a terreno para celebrar el Día Internacional de la Mujer. *El Mercurio*, 8 March 2013.

El Mercurio (2013b). Raúl Gamarra, El origen del boom gastronómico. *El Mercurio*, 7 June 2013.

El Mercurio (2013c). Presidente Piñera encabezará consejo de gabinete para analizar últimos meses de gobierno. *El Mercurio*, 25 October 2013.

El Mercurio (2014a). Marcela Osorio retorna hoy a las teleseries interpretando a una esforzada mujer de La Vega. *El Mercurio*, 5 May 2014.

El Mercurio (2014b). Los Chinganeros, Cuecas de barrios populares. *El Mercurio*, 2 July 2014.

Fincher, R. and Iveson, K. (2008). *Planning and Diversity in the City: Redistribution, Recognition and Encounter*. Houndmills, Basingstoke, UK and New York: Palgrave Macmillan.

Gobierno Regional de Santiago (2014). *Elaboración de un Plan Maestro de Regeneración para el Barrio Mapocho La Chimba*. Unpublished report. Available from: https://www.gobiernosantiago.cl/wp-content/uploads/2014/doc/estudios/Estudio_Elaboracion_de_un_Plan_Maestro_de_Regeneracion_para_el_Barrio_Mapocho_La_Chimba,_2012.pdf (accessed 2 March 2017).

Guárdia, M. and Oyón, J.L. (2015). *Making Cities through Market Halls. Europe, 19th and 20th Centuries*. Barcelona: Museu D'Història de Barcelona.

González, S. and Waley, P. (2012). Traditional retail markets: The new gentrification frontier? *Antipode*, 45(4), 965–983.

Habermas, J. (1990). *Strukturwandel der Öffentlichkeit. Untersuchungen zu einer Kategorie der bürgerlichen Gesellschaft*. Frankfurt: Suhrkamp.

IEUT and CCHC (Instituto de EstudiosUrbanos y Territoriales and CámaraChilena de la Construcción) (2016). *ICVU 2016 Indicador de Calidad de Vida de Ciudades Chilenas*. Santiago: Instituto de Estudios Urbanos y Territoriales and Cámara Chilena de la Construcción. Santiago.

I.M. de Independencia(IlustreMunicipalidad de Independencia) (2014a) *Plan Regulador Comunal*. Santiago: Ilustre Municipalidad de Independencia.

I.M. de Independencia(IlustreMunicipalidad de Independencia) (2014b). *Ordenanza del Plan Regulador de la Comuna de Independencia*. Santiago: Ilustre Municipalidad de Independencia.

I.M. de Recoleta(IlustreMunicipalidad de Recoleta) (2006). *Plan Regulador Comunal*. Santiago: Ilustre Municipalidad de Recoleta.

I.M. de Santiago (IlustreMunicipalidad de Santiago) (1961). *Centros de abastecimiento de barrios: adecuación de ferias libre. Plan General urbano de Santiago: renovación urbana organización de ferias de barrio.* Santiago: D.O.M. (Dirección de Obras Municipales).

I.M. de Santiago (IlustreMunicipalidad de Santiago) (2016). *Explanada de los Mercados – Concurso Público.* Available from: http://www.munistgo.info/concursoexplanada/

La Tercera (2013). Aumento de clientes ABC1 impulsa modernización de la Vega Central. *La Tercera,* 24 June 2013.

Lees, L., Slater, T. and Wyly, E. (2008). *Gentrification.* London and New York: Routledge.

Lefevbre, H. (1969). *El derecho a la ciudad* ['*The Right to the City*'] Barcelona: Edición 62.

López-Morales, E. (2011). Gentrification by ground rent dispossession: The shadows cast by large-scale urban renewal in Santiago de Chile. *International Journal of Urban and Regional Research,* 35(2), 330–357.

Marcuse, P. (1985). Gentrification, abandonment and displacement: Connections, causes and policy responses in New York City. *Urban Law Annual; Journal of Urban and Contemporary Law,* 28, 195–240.

Marcuse, P. (2009). From critical urban theory to the right to the city. *City,* 13(2–3), June–September 2009, 185–197.

ODEPA (2002). *Estudio: Los supermercados en la distribución alimentaria y su impacto sobre el sistema agroalimentario nacional.* Santiago: Oficina de Estudios y Políticas Agrarias del Ministerio de Agricultura, ODEPA.

Pádua Carrieri, A., Dutra Murta, I., Teixeira, J., Machado, B. and Tijoux, Ma E. (2012). Metamorfoseando los mercados centrales. El turismo gastronómico como estrategia en el Mercado Central de Santiago (Chile) y el Mercado Municipal de São Paulo (Brasil). *Estudios y Perspectivas en Turismo,* 21(1), 88–107.

Pujol, J. (2016). *Estudio Mapocho – La Chimba y concurso explanada de los mercados.* Presented at Bienal de Patrimonio, miércoles 18 May 2016. Available from: http://www.achm.cl/index.php/capacitaciones/item/download/122_8b4fac42e4a08a3cce f9eb37af1ebe94

Salazar, G. (2003). *Ferias Libres: Espacio residual de soberanía ciudadana (Reivindicación histórica).* Santiago: Ediciones SUR.

Selle, K. (2003). Was ist los mit den öffentlichen Räumen. *Informes AGB* 49(2). Aachen, Dortmund, Hannover: Dortmunder Vertrieb für Planungsliteratur.

Servicio de ImpuestosInternos (2014). *Rol Semestral de contribuciones bienes raíces predios no agrícolas.* Santiago: Servicio de Impuestos Internos Norte.

Siebel, W. (2000). Wesen und Zukunft der europäiscehn Stady. *DISP,* 141, 28–34.

Siebel, W. (2007). Vom Wandel des öffentlichen Raumes. In: Wehrheim, J. (Ed.), *Shopping Malls: Interdisziplinäre Betrachtungen Eines Neuen Raumtyps (Stadt, Raum und Gesellschaft).* Wiesbaden: VS Verlag für Sozialwissenschaften, pp. 77–94.

Skinner, C. (2008). The struggle for the streets: Processes of exclusion and inclusion of street traders in Durban, South Africa. *Development Southern Africa,* 25(2), 227–242.

The Daily Meal website (n.d.). 45 Best Markets Around the World. Available from: http://www.thedailymeal.com/45-best-markets-around-world-slideshow (accessed 26 October 2014).

Valentine, G. (2008). Living with difference: Reflections on geographies of encounters. *Progress in Human Geography,* 32, 323–337.

Wehrheim, J. (Ed.) (2007). *Shopping Malls: Interdisziplinäre Betrachtungen Eines Neuen Raumtyps (Stadt, Raum und Gesellschaft).* Wiesbaden: VS Verlag für Sozialwissenschaften.

Wehrheim, J. (2015). El caracter público de los espacios y de la ciudad. Indicadores y reflexiones para el posterior desarrollo del tema. In: Schlack, E. (Ed.), *POPS, el uso público en espacios privados*. Santiago: Ediciones ARQ, pp. 287–308.

Weber, M. (1964 [1922]). *Economía y sociedad*, tomo II. México: Fondo de Cultura Económica.

Zukin, S. (2010). *Naked City. The Death and Life of Authentic Urban Places*. New York: Oxford University Press.

Zukin, S., Trujillo, V., Frase, P., Jackson, D., Recuber, T. and Walker, A. (2009). New retail capital and neighborhood change: Boutiques and gentrification in New York City. *City & Community*, 8(1), 47–64.

4 Resisting gentrification in traditional public markets

Lessons from London

Sara González and Gloria Dawson

Introduction

This chapter discusses traditional retail markets in London as spaces where traders and citizens mobilise around wider urban and political issues. Recently in London campaigns have emerged to defend traditional markets from various threats: privatisation, gentrification, closure, demolition, displacement of traders, rent hikes or abandonment and disinvestment (González and Dawson, 2015). These struggles exist in a city that is increasingly unaffordable for low- and middle-income residents while at the same time being a profitable real estate market for global investment companies (Beswick et al., 2016). In particular, housing has become a key space for contestation, with many campaigns fighting against public housing demolition or privatisation, evictions or against the development of luxury and speculative housing projects (Lees and Ferreri, 2016; Watt and Minton, 2016). Less attention has been paid to other spaces where political mobilisation against the neoliberalisation of the city is taking place, such as work or retail spaces (although see Just Space, 2015).

In this chapter, we focus on three inner London markets and campaigns to protect or save them; Queen's Market, Seven Sisters Market and Shepherd's Bush Market (see Figure 4.1). They are all undergoing or have recently undergone processes of redevelopment which threaten their sustainability. They are all situated in relatively deprived neighbourhoods and serve an ethnically-mixed and generally low-income population. In all three cases there has been a period of neglect and disinvestment which has then been used as partial justification for an intervention that would see the markets' physical infrastructure upgraded, but also, crucially, the social mix of the traders and customers changed – towards a higher-income and higher-value goods profile.

Although this chapter focuses on London, market campaigns have been emerging across the UK in the past few decades (for an account, see González and Dawson, 2015). However, this phenomenon is largely invisible in the academic literature and more generally in the national media (although see various articles in *The Guardian*, e.g. Dobson, 2015 and Perry, 2015). These campaigns take various shapes, sometimes involving only traders, sometimes only market users and many times bringing both together. At times they focus on specific

Figure 4.1 Location of markets discussed in this chapter
Source: Annotated Google Map.

demands regarding a particular market, while other times they engage with wider changes in the borough and/or city, forming alliances with other groups. As we will see, in these struggles, the market becomes a metaphor for the city where issues of urban justice take a particular form. In particular, in this chapter we will interpret these 'market struggles' as anti-gentrification mobilisations. The chapter also aims to expand our understanding of gentrification and anti-gentrification beyond residential struggles to incorporate what can be interpreted as 'retail gentrification'. Finally, we also want to discuss the relevance of traditional retail markets as political spaces for mobilisation around citizenship and the right to the city.

This work is based on action research carried out with groups new to us and campaigns and networks in which we were already engaged. It was mostly funded by an Antipode Scholar-Activist award (2014–2015).[1] Through initial desk research 10 key campaigns were identified, three of which are drawn on in this chapter. Initial online contact developed into visits and meetings. As embedded activist researchers, our focus was not so much on producing verbatim accounts of interviews or objective data analysis as on reflecting, organising and deepening collective knowledge and power. Towards the end of the project, a report was published (González and Dawson, 2015) and disseminated widely with market campaigns and policy-makers from key national organisations. The research period culminated in a meeting in January 2016 in London, involving of a group of individuals involved in or interested in market campaigning, including campaign representatives, food activists and researchers. From this an informal network was formed to share information and support.

The following section of the chapter outlines markets as new frontiers of gentrification and contestation by first outlining 'retail gentrification' as distinct from gentrification more generally. The next three sections focus on London, taking first historical issues surrounding London markets in terms of legal frameworks, planning and history, moving to what we identify as a moment of transition for the capital's traditional markets. Following this, three case studies of recent London market campaigns are described. We then draw out some of the wider processes that are challenged by these campaigns, including property speculation, rent rises and displacement of users and traders. The penultimate section focuses on tools and tactics mobilised by these groups. We briefly conclude the chapter with the main arguments returning briefly to the theoretical premises set out at the start.

Markets as new frontiers of gentrification and contestation

Gentrification is now an established concept across the international academic community of critical urban studies, and in activism around urban issues. Using Clark's (2005, p. 263) broad definition, we understand gentrification as

> a process involving a change in the population of land users such that the new users are of a higher socio-economic status than the previous users, together with an associated change in the built environment through a reinvestment in fixed capital.

Gentrification inevitably involves displacement of some residents and users of urban space by wealthier and higher-income users. Some authors argue that gentrification has become a global trend, present simultaneously in many different cities, albeit taking many different forms according to diverse geographical contexts (Lees et al., 2016). Traditionally, gentrification has been analysed in relationship to housing, looking at residential and neighbourhood changes and policies which involve the displacement of working-class and lower-income populations. However, gentrification also takes place in retail spaces and although this has been previously neglected in the critical urban studies literature, there is now emerging evidence of this process. We define retail gentrification as:

> the process whereby the commerce that serves (amongst others) a population of low income is transformed/replaced into/by a type of retail targeted at wealthier people. From a different angle, we can also see it as the increase in commercial rents that pushes traders [and retailers] to increase the price of their products, change products or change location.
>
> (González and Dawson, 2015, p. 19)

Zukin et al. (2009) have discussed the increase in boutiques and large retail chains in some neighbourhoods in New York at the expense of local traditional stores. Similarly, research shows how residents have been displaced in a neighbourhood in Santiago de Chile by trendy shops and bars (Schlack and

Turnbull, 2016), and in a heritage-rich neighbourhood in Shanghai (González Martínez, 2016). As we already discussed in the introduction to this book, gentrification is also taking place in traditional public markets due to multiple processes coming together: decline of the traditional forms of shopping with the explosion of globalised retail businesses such as supermarkets; revalorisation of the heritage value of market buildings; trends of commodification and 'touristification' of food; local and public authorities underinvesting in their retail assets while at the same time wanting to maximise their real estate values. Markets, therefore, are becoming new frontiers for gentrification processes. This is inevitably bringing about conflict between different publics, communities and classes as well as about what role markets should fulfil in cities.

If the literature on retail gentrification is still limited, research and analysis on processes resisting this gentrification is even scarcer and scattered across various areas. Some of the work focuses on grassroots mobilisations against supermarkets and retail chains: In the US, Sites (2007) discusses a campaign against the opening of the retail chain Walmart because of their poor treatment of staff. In the UK there have been many campaigns against the opening of Tesco stores (a large supermarket chain) (Clement, 2012). In Istanbul the Gezi park mobilisation in 2013 originally started as a protest against the opening of a shopping centre (Kuymulu, 2013). In the global South, as discussed in the introduction of this book, one of the most visible and violent expressions of gentrification is the forced eviction of informal traders and customers from central urban areas, processes which are generally strongly resisted by organised traders (Bromley and Mackie, 2009; Brown et al., 2010; Cross, 2000; Crossa, 2009).

Markets and retail spaces therefore are increasingly becoming not only frontiers for gentrification but also for new struggles against it (González and Waley, 2013). Various chapters in this book report on such struggles, as well as further work from Latin America (Boldrini and Malizia, 2014; Delgadillo, 2016). In contrast to contestation around housing issues, these mobilisations involve people who work *and* use/consume in these spaces and who are demanding a right to the city in different ways. The importance to add retail to struggles for the right to the city has been in fact highlighted by Zukin et al. (2009, p. 62) when asserted that 'the right to the city passes through the right to shop there'; and, we would also add (especially in the context of small businesses), to work there too.

Changes and challenges for London market trading in a historical and contemporary context

Markets have had a strong presence in the streets in London, emerging in many different forms. In the second half of the seventeenth century, there was a rise in formally recognised markets (Smith, 2002) and by late eighteenth century, London had over 30 such markets. However, by the nineteenth century, formal markets struggled to expand at the same rate as the population and demand. Unlike in other major cities such as Barcelona or Paris, in London there was no centrally organised plan for the expansion of municipal markets (Fava et al., 2016; Jones, 2015). To cover that gap, informal and unregulated street markets emerged as the alternative. Initially, costermongers, itinerant traders, bought goods from wholesale markets and moved around selling in mobile stalls. Later they started to trade in fixed locations (Kelley, 2015) partly in response to their repression and stigmatisation by the public authorities (Jones, 2016; Kelley, 2015). By the beginning of the twentieth century there were around 8,000 stalls on street markets in London, the majority located in the poorest neighbourhoods (Kelley, 2015). Jones (2016, p. 71) explains their social value:

> Costermongers' flexibility, sustainability and efficiency meant that they could rapidly find buyers for produce that had been designated as wastage by others. Traders had low overheads and were able to sell smaller amounts of inexpensive goods to customers who could not afford to buy larger amounts of higher-quality produce elsewhere.

Today, London still has many markets which play a vital role in the life of Londoners, particularly the poorest ones. A recent report records 99 markets in central and inner London (Cross River Partnership, 2014) and a slightly earlier report on the whole of London reported 162 (Regeneris, 2010). In inner and central London, the turnover of markets in 2014 is estimated at £360m per annum (Cross River Partnership, 2014). A range of policy reports indicate that many markets in London are in a moment of transition. There has been a growth in the number and turnover of markets but this is mainly amongst privately-run and owned markets and more 'niche' markets catering for a wealthier clientele: farmers' markets, speciality markets, street food, craft markets, etc (ibid.). There is also a trend for municipal markets to switch to being managed and/or owned by private operators. From 2008–2014 there was an increase of 9 per cent in private markets in London from 30 to 39, and a decline of local authority-run markets from 70 per cent to 54 per cent (Cross River Partnership, 2014). The *Financial Times* reports that farmers' markets are part of the mix that attracts wealthy residents to expensive central neighbourhoods in London (Cox, 2015). The model markets that are often signalled in policy literature and media as successful are all either tourist destinations or mainly target high-income customers. The more traditional markets, mostly run by local authorities, seem to do less well and these are also more likely to be located in deprived communities (Regeneris, 2010). To explain the decline of the traditional London market, reports highlight changing consumption patterns (the rise of the internet, and supermarkets, especially small ones) but also lack of investment by local authorities (Cross River Partnership, 2014; Regeneris, 2010). Additionally, many London markets are in key central locations and experience pressure from local authorities and developers to be displaced to realise the high land values by building something more profitable instead (González and Dawson, 2015).

As in the past, these reports also show clearly that markets continue to be important for the poorest and most vulnerable communities. There is a clear relationship between the spatial distribution of markets and areas of deprivation in inner and central London; there is also a correlation between the location of markets and those areas with a highest number of Black and minority ethnic (BME) populations, who tend to have lower incomes (Cross River Partnership, 2014). Markets in London also showcase the ethnic and cultural diversity of the city and there has been some research exploring how markets improve communication and understanding between diverse groups (Dines, 2007; Watson, 2009). They also are particularly important for providing access to affordable, accessible fresh food. They also act as 'meeting places and locations for social exchanges, for learning about food and for engaging in the community. The benefits appear to be particularly important for the elderly' (NEF, 2005, p. 54). This chapter focuses on several such markets.

To understand the public discourse around the 'decline' of traditional markets in the London and to analyse the campaigns and mobilisations around them, we need to put the phenomenon in the context of wider changes in global retail and also urban development and urban policies. The decline and renaissance of markets in London occurs within a context of regeneration, state-led gentrification and austerity urbanism (Watt and Minton, 2015). London is experiencing many forms and levels of gentrification, with low-income residents struggling to afford to live, work and even shop in central spaces. Financialisation of the UK economy has particularly affected housing in the capital, as London property is increasingly seen as an investment not only for individuals but also for the state (Edwards, 2016; Watt and Minton, 2016). In some neighbourhoods where our case study markets are based, the population living in social housing peaked at 82 per cent in the 1980s (Watt and Minton, 2015). But much purpose-built social housing is now undergoing massive processes of regeneration, with large-scale demolition, privatisation and rebuilding of new housing aimed at middle class residents. The result is the displacement of many low-income residents from these central areas (London Tenants Federation, 2014), which is not uncontested (Lees and Ferreri, 2016). Public markets, like housing, have become a frontier for gentrification. They are regarded by public and private actors as under-realised opportunities, where a higher profit uses could be developed.

Three market campaigns in London

In this context of regeneration and gentrification, a number of campaigns to protect public markets in London have emerged. This chapter focuses on three campaign groups in London. We describe the emergence and configuration of these groups; in the next section we also analyse their tactics and campaign tools used as well as how the concept of the public market has been explored in each case.

Queen's Market is owned and managed by Newham Council, one of the 32 councils in the Greater London Authority. The market, in northeast London, sits in one of the poorest London neighbourhoods with high ill-heath and one of the highest unemployment rates in the city. Even by London standards it has a high BME population (London's Poverty Profile, 2015). Queen's Market is a key local resource for the affordability and variety of foods, including for those BME groups. A market has existed for many years but the current canopy building dates from 1963 (Percival, 2009). In 2003 traders and customers found out that the market was up for sale, and in 2004 Newham council revealed that it had partnered with a private developer, St Modwen, for a regeneration plan that involved the demolition of the existing market, the construction of a residential tower block, an ASDA supermarket and a much smaller market. According to the council and developers this regeneration was needed because the market had 'reached the end of its useful life' and 'traders are working in the dark ages' given the lack of hygiene facilities (Percival, 2009, p. 31). These plans attracted opposition from the local community and Friends of Queen's Market (FoQM) was born. In 2006 a petition was submitted against the plans with 12,000 signatures; over 2,500 separate objection letters were also sent. A revised planning application for the development was approved by the council, although after much lobbying from campaigners it was revoked by London's Mayor, Boris Johnson, in 2009. Queen's Market was saved from demolition and a redevelopment. However, FoQM has carried on campaigning, as they believe the market is under permanent threat; in 2011 Newham Council earmarked it as 'strategic site' for a 'mixed used development' (FoQM website, n.d.). This was contested by FoQM and eventually the threat was removed. However, the neighbourhood is experiencing other forms of gentrification and the market suffers a general state of disrepair and disinvestment.

In Seven Sisters, North London, the mainly Latin American indoor market of around 40 shops and stalls has been under threat of demolition and relocation since 2007. Traders and the wider community have been fighting through legal challenges and complex campaigning. The Market is located in Haringey, which, like Newham, is one of the poorest neighbourhoods in London and in the UK, with a very diverse population (Haringey London, 2015).

In 2004, the local authority started negotiations with the developer Grainger to redevelop the whole area, named Wards Corner, with the market at its heart. The proposals, crafted without proper consultation with residents and local businesses, involved the demolition of the indoor market (Román Velázquez, 2013), the building of private flats in a gated style (WCCC, 2008) and demolition of local architectural heritage. By late 2007 a community group called Wards Corner Community Coalition (WCCC) had formed to oppose the plans and propose alternatives, with support from residents' associations, the market traders' association and heritage and local cultural associations (WCCC, n.d. a and b). In March 2008 a planning application was submitted by the developer, Grainger, which was initially approved by the public authority. However, WCCC challenged this in the courts on the basis that it did not consider the negative impact on the Latin American community using the market; this resulted in the planning application being quashed in 2010 (WCCC, 2012). Eventually the developer acquired planning permission for a mixed-use project involving new retail space and 196 non-'affordable' housing units. In October 2016, Haringey Council brought forward its power to compulsorily purchase properties in the area to facilitate the developer's land assembly which WCCC and traders opposed. The project is now on hold, subject to a public inquiry by the Secretary of State of local government (Haringey London, 2016). In parallel to the developers' plan, WCCC, other campaigners and traders have been developing an alternative community plan for which they acquired planning permission in 2014 (WCCC, 2014a), have set up a Development Trust and have been linking with other organisations in the area to campaign on issues centring on the best interests of diverse local communities (Our Tottenham, n.d.).

In West London, the Shepherd's Bush Market Tenants' Association (SBMTA) have been fighting for years against a regeneration project which would have transformed the area around the 100-year-old market. SBMTA is an independent association, representing the vast majority of 140-odd market businesses. The market is a partly street and partly covered market stretching along a train track. Again, this market is situated in a highly diverse neighbourhood (Hammersmith and Fulham, 2011). It is also one of the most deprived areas in the highly polarised borough of Hammersmith and Fulham, with lower than average household incomes and academic qualifications (ibid.). A survey conducted in 2008 showed that the market relied heavily on low-income population groups (GVA Grimley, 2008). The report also showed the strong presence of ethnic minority users.

Like Queen's Market, the proposed private development in and around Shepherd's Bush Market would have created around 200 residential units, none of which would be 'affordable'. It would also include refurbishing the market. The local authority and the private developers have both suggested the need to 'enhanc[e] the Market's offer with a more diverse mix, complemented by new retail, café and restaurant uses' (Hammersmith and Fulham, 2011: section 4). Traders have been concerned that although in theory the new development would keep the market, it would vastly change the area and bring new residents with different consumption preferences unlikely to prefer existing businesses. Even before this proposed development, traders denounced the lack of investment in the market by the previous owner, Transport for London (TfL), a significant London landlord (Horada, 2013). In 2011 a private developer, Orion, in partnership with Development Securities PLC, bought a large part of the market and adjacent plots to enable their development. Their initial planning application was not approved by the Mayor of London as it did not justify the lack of affordable housing and was unclear on how the 'unique character' of the market would be maintained (Greater London Authority, 2011). Eventually the scheme was given planning permission by the local authority in March 2012. However, a series of legal challenges led by traders and independent business owners and a public inquiry in 2014 by the relevant Secretary of State have challenged the project repeatedly and finally in September 2016 the developers announced that they would no longer pursue the project (Prynn and Diebelius, 2016). Despite this victory, the long dispute and uncertainty has had a significant impact on the market and around 10 per cent of traders left the market in 2014–2015 (Leigh Day, 2015).

In this section we have explored three examples of long-term campaigns by London market traders and users. In the next section we provide a more detailed analysis of how we can understand these contestations through the lens of gentrification and how traders and campaigners are mobilising around markets not just as work or retail spaces but also as a community resources.

Gentrification and displacement in traditional retail markets in London

We see the three campaigns described in the previous section (and several others currently active in London) as campaigns against the gentrification of community and workspaces directly challenging five important processes linked to gentrification: 1) Abandonment and disinvestment; 2) Property speculation, 3) Plans to 'upscale' their socioeconomic profile; 4) Higher rents and prices and 5) Potential displacement of certain users and traders. Using our case studies, we illustrate these processes in turn.

The three market campaigns are struggling against abandonment and/or disinvestment by managers and owners. Shepherd's Bush and Seven Sisters were both previously owned (at least in part) by Transport for London (TfL), who increasingly consider their assets as real estate development opportunities (Future of London, 2014). In the case of Shepherd's Bush Market, TfL's share was sold to developer Orion after a long period of neglect. According to local Member of Parliament, Andy Slaughter, deliberate lack of investment by TfL was 'being used as an excuse for demolition of the whole site' (Slaughter, 2013, p. 2). Similarly, in Seven Sisters, WCCC argues that 'Transport for London has allowed their properties [the market and the buildings] to lie derelict and under-occupied despite interest in the buildings from many businesses' (WCCC, n.d. b). Queen's Market is owned and run by a local authority which has neglected investment in the market, which campaigners have repeatedly denounced (see Figure 4.2).

Of course, disinvestment and neglect are often preludes to speculative real estate developments. The three markets are in key urban development areas, enlarging the potential rent gap. The historical neglect of these infrastructures clearly becomes a justification for redevelopment and residents and traders often prefer to support any kind of investment to halt the neglect.

The pressures for 'upscaling' the socioeconomic profile of traders and users in these markets come in different forms. Cash-strapped public authorities and

Figure 4.2 Friends of Queen's Market demonstrating against the lack of investment in the market's infrastructure.
Source: Colin O'Brien.

private developers see the low-income profile of the customers of these markets as an opportunity cost in relation to other potentially more profitable potential uses of the land. They speculate that if higher-income customers were attracted, then traders could upscale their businesses and make more profit which could be captured through higher rents. A report by a commercial property agency concerning Shepherd's Bush market recommends, 'extend[ing] the mix and range of goods to draw in *and retain* expenditure from higher disposable income groups' (GVA Grimley, 2008, p. 12). The private developer that pushed the regeneration project described the current markets as 'underutilised' (Savills, 2011, p. 7).

Displacement of users and traders in these markets takes places in the various ways evidenced in Marcuse's (1985) work on residential gentrification.[2] In the case of retail gentrification, we can identify direct and indirect displacement of market traders, users, products and practices. Direct displacement of traders takes place when markets are demolished without a replacement, but also even when redevelopment projects consider alternative locations or the rebuilding of a market. Sometimes businesses do not survive the disruption of the redevelopment period or cannot find suitable trading places in the higher-rent redeveloped markets. In Seven Sisters, the Latin American market is meant to be moved nearby but according to a consultation carried by the council in 2012, 40 per cent of the businesses surveyed said that they were certain their business would *not* operate in the new premises and 40 per cent were uncertain how to continue (AECON, 2016). This is because rents would be significantly higher. Similarly, in Shepherd's Bush the chair of SBMTA feared that 'the redevelopment of the market will result in units limited to potential retailers abler to afford much higher rents than any market traders could possibly justify' (Horada, 2013). Indirect displacement also takes place but is more difficult to evidence. Lack of investment, particularly in the market's infrastructure, can drive traders out as the trading environment deteriorates. Fewer customers are attracted and their businesses are unsustainable. This has affected Shepherd's Bush in particular.

The displacement of traders has an effect on customers and users. Our three case study markets all sell speciality food and goods from Africa, the Caribbean and Latin America. Users of 'traditional' markets often travel there for specific products that are difficult to find elsewhere at low prices, such as particular foods, or speciality fabrics. If traders selling these products are displaced or if they no longer find it sustainable to sell them because the customer base is changing towards wealthier, whiter residents, then these customers stop going to the market, and are themselves displaced.

Customers and market users might also be displaced when the market no longer seems a welcoming space, as discussed by Shaw and Hagemans (2015) in the case of gentrifying neighbourhoods. This loss of sense of place can happen for many different reasons. FoQM denounced proposals for a regenerated Queen's Market, claiming that they would erase the comforting ethnic and cultural diversity of the place (Dines, 2007). Seven Sisters Market has become a solidarity networking point where new Spanish-speaking migrants can find advice through informal help or NGOs situated in the market (AECON, 2016). This slowly-evolved environment would be difficult to replicate in a redeveloped market with higher rents.

Through an analysis of three markets, we have evidenced the processes of gentrification and displacement taking place in spaces which so far have been neglected in the gentrification literature. In the next section, we discuss how the three campaigns are resisting these processes and how through these tactics they invoke markets as alternative spaces to the gentrifying and speculative city.

Markets as spaces of anti-gentrification resistance

In the section above we have used the framework of gentrification as a critical analytical lens to understand the processes involved in campaigns to protect, save or promote traditional retail markets in London. Although the campaigns do not necessarily always use an explicit gentrification language, their actions can be interpreted as practices of resistance against gentrification as they tackle the challenges described in the previous section. There are various dimensions to this resistance which vary according to the type of campaigns, market and the type of strategies used by the traders and campaigners which we discuss below.

FoQM is led by market users who have been meeting regularly about once a month for more than a decade, with backing from traders and with links to many local organisations and campaigns across London. Campaigning in

Seven Sisters has also been long-running and is more fragmented, with traders and various local organisations working within WCCC and more recently through the N15 Development Trust. In Shepherd's Bush, the campaign is more recent and mainly involves traders with support from local residents. Campaign demands and strategies vary according to these different types of groups. FoQM and WCCC have developed a long-term critique of regeneration in their area, linking with housing and more explicit anti-gentrification campaigns and situating their markets within wider urban struggles. Shepherd's Bush traders are far more focused on the market and its trading conditions.

The three campaigns have been involved in legal and planning contestations to stop developments threatening their markets. Legal appeals are common in urban struggles (see also Lees and Ferreri, 2016) as they can sometimes be the last resort. The kind of gentrification framework that we have explained in the previous section is difficult to translate into the UK legal and planning framework. Thus engaging in these battles has demanded that campaigns use technical arguments, which require expensive solicitors and experts to channel their concerns into appropriate language. In all three cases, the campaigns have challenged the way local authorities have consulted the public about the redevelopment projects, pointing out that the potential negative consequences were not fully explored. WCCC set a precedent in planning terms by challenging the private developers' project over its negative impact on a particular ethnic minority community (Latin American) using the Equality Act 2010.[3] In the three cases, the campaigns have appealed to the special planning powers of the Mayor of London with mixed results. While legal and planning appeals can be restrictive, they also offer a public platform for a wider discussion of explicitly social justice issues such as displacement, gentrification and marginalisation. For example, the planning inquiry over Shepherd's Bush Market was held over eight days, and considered hundreds of documents and heard many of the traders' and residents' views.

FoQM and WCCC, however, have gone far beyond legal and technical battles and have used other strategies to connect with users of the market and other local and London-wide groups. They organise public meetings, protests, stunts, photo calls with the media and participate in London-wide planning consultations. Since 2012, FoQM have been involved in the regular trader-management meetings, where they press on issues arising from traders and market users although they have been banned from these for unclear reasons (FoQM website). Beyond their specific markets, FoQM has been key in the emergence and sustaining of a parallel campaign demanding 100 per cent social housing in a nearby development project (Murphy-Bates, 2015) and WCCC is well networked with other campaigns in the wider neighbourhood. Its Latin American traders have linked with similar campaigns over retail spaces in Elephant and Castle, South London (Román Velázquez and Hill, 2016).

Notably, FoQM and WCCC have patiently developed plans for the markets which express alternative well-researched proposals for their improvement and flourishing. These have been drafted in consultation with traders and the public through participatory planning, sometimes with support from planning experts, architects and academics (Creative Citizens, n.d.). Like many similar campaigns, FoQM has been successful in listing their market an 'asset of community value' which means that the local authority recognises it as 'a place of resort and social interaction for and provides social services [which] furthers the social wellbeing and social interests of the local community' (London Borough of Newham, 2015). As mentioned earlier, WCCC have gone even further, forming a development trust and acquiring planning permission for the development of an alternative project in the site of the Market (West Green Road / Seven Sisters Development Trust, n.d.).

Through these techniques and initiatives the three campaigns have so far managed to stop or delay attempts to redevelop and potentially gentrify their markets, displacing traders and customers. Although these might seem small drops in London's rapid change in the opposite direction, they are important in showing that community and grassroots movements can have an important influence in shaping their urban environments and limiting gentrification.

Indeed, these campaigns have not only resisted and limited the gentrification of the particular retail spaces they work in, they have taken steps to re-inscribe them under a different framework. Instead of looking at markets as mere commercial spaces or real estate assets, they are reclaiming them as public and community spaces, affordable and welcoming for low income and ethnic minority groups. For example, FoQM campaigned successfully to have a special policy on markets in the Mayor's London Plan in 2010 which pro-posed that 'in considering proposals for redevelopment [the Mayor should] consider whether this will impact on economically hard-pressed groups; enable[s] mechanisms to protect the levels of rent necessary for the market's social and locally affordable function' (Friends of Queens Market, n.d.). This policy was later incorporated into a 'community-led plan for London' (Just Space, 2016a) by the campaign network Just Space which has been recently used to highlight the lack of attention in the Mayor's plan over such com-munity work spaces (Just Space, 2016b). Similarly, WCCC has engaged in a public consultation process to provide further alterations to the London Plan in 2014, highlighting the importance of their market as community space and stressing the gap in the planning system that does not recognise the cultural and social value of such diverse spaces (WCCC, 2014b).

Conclusions

In this chapter we have outlined a grassroots resistance to the disappearance, neglect and gentrification of traditional public markets. We have concentrated on three initiatives in London which campaign around three markets in deprived and multi-ethnic neighbourhoods. Markets in London, as well as in other parts of the world, are undergoing a process of transformation. While some types of markets that appeal to high-income consumers are flourishing, the more traditional market which serves vulnerable communities such as low income or ethnic minorities are increasingly under threat. In London, this struggle is exacerbated by high land values where local authorities hit by austerity policies are reconsidering redeveloping them into something more profitable or simply selling them off. Similarly, private developers are looking at the real estate opportunities of central London areas where often-neglected markets are in the way of new retail developments. Gentrification literature in critical urban studies has generally focused on residential transformations and the displacement of vulnerable and working class communities but much less is known about the gentrification and displacement in the retail environment.

This chapter has shown that traditional public markets are affected by gentrification processes; this often starts with neglect and abandonment of the physical infrastructure of the market which later justifies a redevelopment project. These projects, as we have shown in our three cases, sometimes plan to eliminate markets but often aim at upgrading them, not only through invest-ment in the infrastructure but also by 'upscaling' the trader and customer mix. As we have seen this can have the effect of displacing traders who cannot afford the rents in the new redeveloped market and indirectly displaces market users who can no longer afford the produce or feel out of place. Although the market campaigns analysed in this chapter might not use the language of gentrification, they all address these issues of displacement and real estate speculation and will sometimes use terms such as 'ethnic cleansing' or 'social cleansing'. We noted that these campaigns are for the most part networked with other local campaigns and struggles who are making wide claims about the unaffordability of housing and the lack of community spaces in their neighbourhoods. In fact, campaigns mobilise markets not just as commercial spaces or independent businesses but also as community hubs where vulnerable groups can find support and solidarity. These struggles for markets can therefore be understood as the wider fight for the 'right to the city'.

Notes

1 Sara González was awarded a scholar activist award by the Antipode Foundation in 2014. The funding allowed Gloria Dawson to be employed to carry out the research. The project also funded meetings and workshops. See Antipode Foundation website: https://antipodefoundation.org/scholar-activist-project-awards/201314-recipients/
2 Marcuse identified various forms of residential displacement linked to gentrification and abandonment of neighbourhoods: direct displacement, indirect displacement and exclusionary displacement. See Marcuse (1985). See also see Schlack et al., in this book, for an elaboration of Marcuse's displacement discussion adopted to markets.
3 However, the negative impact of a particular project over low-income and econom-ically vulnerable groups of people is not legally recognised as a reason to challenge a planning application in the UK and therefore this cannot be used by campaigns.

References

AECON (2016). Wards Corner Compulsory Purchase Order (CPO) equality impact assessment for Haringey Council. Available from: http://www.minutes.haringey.gov.uk/mgConvert2PDF.aspx?ID=81677

Beswick, J., Alexandri, G., Byrne, M., Vives-Miró, S.Fields, D., Hodkinson, S. and Janoschka, M. (2016) Speculating on London's housing future. *City*, 20(2), 321–341, DOI: doi:10.1080/13604813.2016.1145946

Boldrini, P. and Malizia, M. (2014). Procesos de gentrificación y contragentrificación. Los mercados de Abasto y del Norte en el Gran San Miguel de Tucumán (noroeste argentino). *Revista Invi*, 29(81), 157–191.

Bromley, R.D. and Mackie, P.K. (2009). Displacement and the new spaces for informal trade in the Latin American city centre. *Urban Studies*, 46(7), 1485–1506.

Brown, A., Lyons, M. and Dankoco, I. (2010). Street traders and the emerging spaces for urban voice and citizenship in African cities. *Urban Studies*, 47(3), 666–683.

Clark, E. (2005). The order and simplicity of gentrification: A political challenge. In: Rowland, A. and Bridge, G. (Eds.), *Gentrification in a Global Context: The New Urban Colonialism*. London: Routledge, pp. 261–269.

Clement, M. (2012). Rage against the market: Bristol's Tesco riot. *Race & Class*, 53(3), 81–90.

Cox, H. (2015). Is there a link between London's street markets and house prices? *Financial Times*, 16 October 2015. Available from: https://www.ft.com/content/d8f77e60-70c1-11e5-9b9e-690fdae72044

Creative Citizens (n.d.). Have your say on the community plan for Wards Corner. Available from: https://cc.stickyworld.com/room/presentation?roomid=11#work/122

Cross, J. (2000). Street vendors, and postmodernity: Conflict and compromise in the global economy. *International Journal of Sociology and Social Policy*, 20(1/2), 29–51.

Cross River Partnership (CRP) (2014). Sustainable urban markets: An action plan for London. London: Cross River Partnership. Available from: http://crossriverpartner ship.org/media/2014/12/Sustainable-Urban-Markets-An-Action-Plan-for-London3.pdf

Crossa, V. (2009). Resisting the entrepreneurial city: Street vendors' struggle in Mexico City's historic centre. *International Journal of Urban and Regional Research*, 33(1), 43–63.

Delgadillo, V. (2016). Presentacion. *Alteridades*, 26(51), 3–11.

Dines, N. (2007). The experience of diversity in an era of urban regeneration: The case of Queen's Market, East London. In EURODIV Conference, Diversity in cities: Visible and invisible walls, 11–12 September 2007, London. Available from: http://ebos.com.cy/susdiv/uploadfiles/ED2007-048.pdf

Dobson, J. (2015). Saving our city centres, one local market at a time. *The Guardian*, 10 July 2015. Available from: https://www.theguardian.com/cities/2015/jul/10/sa ving-city-centre-local-market

Edwards, M. (2016). The housing crisis: Too difficult or a great opportunity? *Soundings*, issue 62, 23–42.

Fava, N., Guàrdia, M. and Oyón, J.L. (2015). Barcelona food retailing and public markets, 1876–1936. *Urban History* [online first], pp.1–22.

Friends of Queens Market (n.d.). Website. Available from: http://www.friendsof queensmarket.org.uk/index2.html

Future of London (2014). Spotlight: Taking stock at TfL – an interview with Graeme Craig. 5 June 2014. Available from: http://www.futureoflondon.org.uk/2014/06/05/sp otlight-taking-stock-at-tfl-an-interview-with-graeme-craig/

González Martínez, P. (2016). Authenticity as a challenge in the transformation of Beijing's urban heritage: The commercial gentrification of the Guozijian historic area. *Cities*, 59, pp. 48–56.

González, S. and Dawson, G. (2015). *Traditional Markets Under Threat: Why It's Happening and What Can Traders and Campaigners Do*. Public report. Available from: http://tradmarketresearch.weebly.com/uploads/4/5/6/7/45677825/traditional_markets_under_threat-_full.pdf

González, S. and Waley, P. (2013). Traditional retail markets: The new gentrification frontier? *Antipode*, 45(4), 965–983.

Greater London Authority (2011). Planning report PDU/2764/01: Shepherd's Bush Market, Uxbridge Road in the London Borough of Hammersmith & Fulham planning application no. 2011/02930/OUT. 17 October 2011 Available from: https://www.london.gov.uk/sites/default/files/PAWS/media_id_186560/shepherd's_bush_market_uxbridge_road_report.pdf

GVA Grimley (2008). Shepherd's Bush Market Regeneration, London Borough of Hammersmith & Fulham FINAL DRAFT. February 2008. Available from: http://www.persona.uk.com/shepherds/deposit-docs/CD53.pdf

Hammersmith and Fulham (2011). *Hammersmith and Fulham Shepherds Bush Market Area Planning Brief for Market and Theatre led Regeneration – Equality Impact Assessment Section 1 – Details of Full Equality Impact Assessment*. Available from: http://democracy.lbhf.gov.uk/documents/s6725/07.4%20Shepherds%20Bush%20Market%20-%20appendix%204.pdf

Haringey London (2015). Figures about Haringey. Population profile of Haringey. Available from: http://www.haringey.gov.uk/social-care-and-health/health/joint-strategic-needs-assessment/figures-about-haringey

Haringey London (2016). Wards Corner Compulsory Purchase Order Updates: Seven Sisters Regeneration. 22 December 2016. Available from: http://www.haringey.gov.uk/planning-and-building-control/planning/major-projects-and-regeneration/seven-sisters-regeneration#update-211216

Horada, J. (2013). Letter to the Secretary of State for Communities and Local Government. London Borough of Hammersmith and Fulham (Shepherd's Bush Market Area) Compulsory Purchase Order 2013. Available from: http://www.persona.uk.com/shepherds/E-OP/OBJ-202-1.pdf

Jones, P.T. (2016). Redressing reform narratives: Victorian London's street markets and the informal supply lines of urban modernity. *The London Journal*, 41(1), 60–81.

Just Space (2015). *"London for All". A Handbook for Community and Small Business Groups Fighting to Retain Workspace for London's Diverse Communities*. London: Just Space. Available from: https://justspacelondon.files.wordpress.com/2015/09/workspacehandbook_lowres.pdf

Just Space (2016a). *Towards a Community-led Plan for London. Policy Directions and Droposals*. London: Just Space. Available from: https://justspacelondon.files.wordpress.com/2013/09/just-space-a4-community-led-london-plan.pdf

Just Space (2016b). A response to the Mayor's document A City for All Londoners. 11 December 2016. Available from: https://justspacelondon.files.wordpress.com/2016/12/just-space-response-to-a-city-for-all-londoners.pdf

Kelley, V. (2015). The streets for the people: London's street markets 1850–1939. *Urban History*, 43(3), 391–411. DOI: doi:10.1017/S0963926815000231

Kuymulu, M. B. (2013). Reclaiming the right to the city: Reflections on the urban uprisings in Turkey. *City*, 17(3), 274–278.

Lees, L. and Ferreri, M. (2016). Resisting gentrification on its final frontiers: Learning from the Heygate Estate in London (1974–2013). *Cities*, 57, 14–24.

Lees, L., Shin, H.B. and Lopez-Morales, E. (Eds.) (2016). *Global Gentrifications: Uneven Development and Displacement*. Bristol: Policy Press.

Leigh Day (2015). Market traders appeal High Court decision in legal battle over Shepherd's Bush Market development. Leigh Day website. 19 August 2015. Available from: https://www.leighday.co.uk/News/2015/August-2015/Market-traders-appeal-High-Court-decision

London Borough of Newham (2015). Notice of determination of community nomination Available from: https://www.newham.gov.uk/Documents/Environment%20and%20planning/ACV-QueensMarket.pdf

London Tenants Federation (2014). *An Anti-gentrification Handbook for Council Estates in London.* Available from: https://southwarknotes.files.wordpress.com/2014/06/staying-put-web-version-low.pdf

London's Poverty Profile (2015). Newham, Poverty Indicators. Available from: http://www.londonspovertyprofile.org.uk/indicators/boroughs/newham/

Marcuse, P. (1985). Gentrification, abandonment, and displacement: Connections, causes, and policy responses in New York City. *Journal of Urban and Contemporary Law*, 28, 195–240.

Murphy-Bates, S. (2015) Campaigners condemn lack of social housing in Boleyn ground plans. *Newham Recorder*, 26 January 2015. Available from: http://www.newhamrecorder.co.uk/home/campaigners_condemn_lack_of_social_housing_in_boleyn_ground_plans_1_3930904

New Economics Foundation (NEF). (2005). *Trading Places: The Local and Economic Impact of Street Produce and Farmers' Markets.* London: NEF.

Our Tottenham (n.d.). Our Tottenham website. Available from: http://ourtottenham.org.uk/

Percival, T. (2009). Commercial gentrification in a global city: The changing nature of retail markets in East London. Unpublished MA dissertation, School of Geography, University of Leeds.

Perry, F. (2015). London's local markets under threat from gentrification – readers' stories. *The Guardian* [online], May 19 2015. Available from: https://www.theguardian.com/cities/2015/may/19/london-local-markets-threat-gentrification-stories-brixton

Prynn, J. and Diebelius, G. (2016). Victory for Shepherd's Bush Market traders as £150m flats plan is axed. *Evening Standard*, 7 September 2016. Available at: http://www.standard.co.uk/news/london/victory-for-market-traders-as-150m-flats-plan-is-axed-a3338836.html

Regeneris Consulting (2010). London's retail street markets: Draft final report. London: Regeneris Consulting Ltd. Available from: https://www.london.gov.uk/file/8012/download?token=Wt9HoCMq

Román Velázquez, P. (2013). Valuing the work of small ethnic retail in London: Latin retail at E&C and Seven Sisters. Event on 'Alternative Strategies for economic development in London'. Just Space and University College London, 23 March 2013. Available from: https://latinelephant.files.wordpress.com/2015/04/valuing-small-ethnic-retail-space_ec.pdf.

Román Velázquez, P. and Hill, N. (2016). The case for London's Latin Quarter: Retention, growth, sustainability. London: Latin Elephant. Available from: http://latinelephant.org/wp-content/uploads/2015/03/The-Case-for-Londons-Latin-Quarter-WEB-FINAL.pdf.

Savills (2011). Shepherd's Bush Market Toolkit Viability Assessment. Available from: https://www.concreteaction.net/wp-content/Documents/Viability/Shepherds-bush-market-viability-assessment

Schlack, E. and Turnbull, N. (2016). Emerging retail gentrification in Santiago de Chile: The case of Italia-Caupolicán. In: Lees, L., Shin, H.B.and López-Morales, E. (Eds.), *Global Gentrifications: Uneven Development and Displacement*. Bristol: Policy Press, pp. 349–373.

Shaw, K.S. and Hagemans, I.W. (2015). 'Gentrification without displacement' and the consequent loss of place: The effects of class transition on low-income residents of secure housing in gentrifying areas. *International Journal of Urban and Regional Research*, 39(2), 323–341.

Sites, W. (2007). Beyond trenches and grassroots? Reflections on urban mobilization, fragmentation, and the anti-Wal-Mart campaign in Chicago. *Environment and Planning A*, 39(11), 2632–2651.

Slaughter, A. (2013). Proof of evidence of Andy Slaughter MP, London Borough of Hammersmith & Fulham (Shepherd's Bush Market Area) Compulsory Purchase Order Public Inquiry 2013. Available from http://www.persona.uk.com/shepherds/E-OP/OBJ-283.pdf

Smith, C. (2002). The wholesale and retail markets of London, 1660–1840. *The Economic History Review*, 55(1), 31–50.

Wards Corner Community Coalition (WCCC) (n.d., a). Who we are. WCCC website. Available from: http://wardscorner.wikispaces.com/-+Who+We+Are

Wards Corner Community Coalition (WCCC) (n.d., b). History. WCCC website. Available from: http://wardscorner.wikispaces.com/History

Wards Corner Community Coalition (WCCC) (2008). 31st March deputation to the full council. WCCC website. Available from: http://wardscorner.wikispaces.com/31st+March+Deputation+to+the+Full+Council

Wards Corner Community Coalition (WCCC) (2012). Judicial review. Available from: https://wardscorner.wikispaces.com/-+Judicial+Review

Wards Corner Community Coalition (WCCC) (2014a). WCCC is granted permission for a community-led plan to restore Seven Sisters Market. Wards Corner Community Plan website. Available from: https://wardscornercommunityplan.wordpress.com/2014/04/26/wcc-is-granted-permission-for-a-community-led-plan-to-restore-seven-sisters-market/

Wards Corner Community Coalition (WCCC) (2014b). Further alterations to the London Plan: Submission from Wards Corner Community Coalition. April 2014. Available from: https://justspacelondon.files.wordpress.com/2014/04/falp-wcc-submission.pdf

Watson, S. (2009). Brief encounters of an unpredictable kind: Everyday multiculturalism in three London street markets. In: Wise, A. and Velayutham, S. (Eds.), *Everyday Multiculturalism*. London: Palgrave Macmillan, pp. 125–139.

Watt, P. and Minton, A. (2016). London's housing crisis and its activisms: Introduction. *City*, 20(2), 204–221.

West Green Road / Seven Sisters Development Trust (n.d.). Website. Available from: https://n15developmenttrust.wordpress.com/

Zukin, S., Trujillo, V., Frase, P., Jackson, D., Recuber, T. and Walker, A. (2009). New retail capital and neighborhood change: Boutiques and gentrification in New York City. *City & Community*, 8(1), 47–64.

5 The contested public space of the *tianguis* street markets of Mexico City[1]

Norma Angélica Gómez Méndez

Introduction

> given that we are in a public space, if it is public, then everyone can have a piece of the action, am I right?
>
> (Alejandro, Leader of trader organisation in Iztapalapa district, Mexico City, 21/11/2009)

Urban space in large Latin American cities, particularly Mexico City, is full of tension, conflict, experiences of solidarity, cooperation and collective action. These take place in a context of policies of structural adjustment and the liberalisation of markets where urban public space becomes increasingly precarious, unequal, deprived and contradictory. Under this premise of contestation by a large number of actors, the understanding of urban public space requires examination of how it is experienced and used and the demands of those that inhabit it with their diverse and divergent positions (Bourdieu, 1999; Hiernaux, 2013).

The trade that takes place on the streets of Mexico City refers precisely to the debate about how public space is inhabited. Various studies of this kind of informal occupation (Bromley, 2000) have discussed not only the pre-Colombian tradition of selling products on the streets, but that currently the space is contested mostly by those who use it as an alternative source of work (this in the context of the shortage of formal employment and the precarious nature of that which exists). However, this alternative source of work implies the occupation of public space for private purposes regulated by a legal framework that did not contemplate its growth, the conditions under which it is carried out, its forms of organisation and collective action, nor the relationship of these organisations with the authorities.

This type of trade has different manifestations that depend on the physical space, the products that are sold and how they are traded (see also Delgadillo, in this book, for a discussion of various forms of retail in Mexico City). They include: *tianguis* or flea markets (a movable semi-fixed space on the street every day of the week); 'concentrations' or market stalls (a fixed or semi-fixed position installed in the same place throughout the week); 'wheeled' markets (with the same spatial dynamics of the *tianguis* but selling different products); and street traders who regularly sell products in public spaces in a nomadic manner and without a permanent position.

This diversity of types of retail translates into very heterogeneous forms of daily work and street trading, where conflicts are not the same in terms of influencing strategies of income, the degree of organisation required to work in the public space and the demand of rights, which at their most basic can be expressed as the right to work and the right to the city.

This chapter explores in detail one of the types of street trade in Mexico City: the *tianguis*. Their importance originates in their pre-Colombian tradition and as supplier of goods to the working class. In addition the *tianguis* are organised as not-for-profit associations registered with the government authorities and are relatively more formal and stable in their occupation of space. This research uses mixed methodologies, including analysis of documents relating to trade in public space in Mexico City, quantitative data prepared by government institutions and fieldwork which includes life-histories of the leaders of the organisations and interviews with the traders of the *tianguis* in Iztapalapa, one of the most marginalised and populous municipalities of the city.

The chapter is organised in two sections: the first defines the public space as a territory of power where actors establish strategies of negotiation and organisation in order to appropriate space – of which commerce in the public realm is an expression – and describes the most important factors and characteristics of this activity; the second analyses the quantitative dimension, internal organisation and political logic of the *tianguis* to demonstrate the ways in which citizens contest and negotiate a place on the public street, expressed in the words of those who work there.

Trade on the streets of Mexico City: The contestation of public space

In the social sciences, a fundamental coincidence can be observed in the debate about public space; i.e. that the experience of public space is characterised by the conflict and coexistence of multiple demands and its social and political actors which generate a constant dispute over the rights of how the space is used and by whom. Ramírez, quoting Sennett, points out that public space 'refers to a wide variety of people representing a cosmopolitan and "multi-form urban public" whose scenario for interaction is the capital city where complex social groups converge' (Sennett, 2011, in Ramírez, 2013, p. 287).

Street traders are social agents in a conflictive environment politicised by the use, distribution and maintenance of a scarce resource, the street. Here many actors come together under the assumption that the State has the monopoly and legal recognition for the administration of the street because it controls the various capital and actors who are involved in other areas (Bourdieu and Wacquant, 2005). Among the most important of these are: the officials and representatives of government who are responsible for implementing the legal framework; the political parties who dispute the space as well as its agents (traders, their organisation and leaders); the formal merchants and neighbours of the *tianguis*. Between these actors, spheres of power and conflict are constructed into which commerce enters and operates. Access to the public street is through the basic elements of leadership and organisations.

For the members of the trade organisations the possibility to occupy public space as individuals is virtually non-existent and it is these social networks, or family connections that are used to facilitate entry into street trade. For the leaders of the organisations their involvement begins with street trade and is maintained over time: very often leadership is extended and consolidated through family ties because control over the use and distribution of space is, in fact, the reason for the existence of the leader. Organisations have arisen through constituting civil associations and at least in Mexico City this collective model was imposed by the State as a requirement to obtain permission to trade on the streets (Cross, 1998). These organisations were also used by traders and leaders to transform themselves into agents capable of generating social demands.[2]

Street trade is only one part of the complex reality of public space where a fundamental issue is put into play; the right to work and the imperative to engage in a functioning economic activity. This can be seen either as a survival strategy for those who have no access to formal work or as a source of additional income where formal work is financially insufficient. On the other hand this street trade contributes to the distribution and supply of products and services, many of which are basic necessities, and functions as a space for socialising and encounters, representing an opportunity to maintain the vitality of a public space.

Bromley (2000) has already indicated some of the arguments in favour of street trade, some of which relate to the discussion of civil rights: the constitutional right to choose an occupation or business activity, with street trading a part of that right. Street trade also contributes to the development of independent work and functions as a safety valve for the unemployed (Cross and Morales, 2007; Morales and Kettles, 2009), because without this type of activity many turn to crime. Sethuraman (1998) and Maloney (2004) state that this trade is an opportunity for family businesses entrepreneurship and provides flexibility of labour, particularly for women by allowing them more financial independence and greater control over their hours of work in spite of the domestic activities that they undertake such as looking after children and family members (particularly older people), cleaning the home, washing clothes and cooking.

For their part Morales and Kettles (2009) recognise the importance of this trade as a means of supply of basic goods combined with the possibility of building more attractive social spaces: 'Public markets and street vendors can be temporary uses or more permanent responses to consumer demand, economic inequality, and mobility-constrained populations. When properly sited, they provide neighbourhood amenities and can contribute to a positive community image' (Morales and Kettles, 2009, p. 2).

Definitions and data: The complexity of street trade in Mexico City

Street trade highlights the dispute over public space between the different actors involved (government, traders, leaders, etc.), often because the trader organisations occupy public space without authorisation or without respecting legal frameworks and do not report the exact number of traders. This means that much of the information is not publicly available, or rather that there is a real 'war of numbers' which makes it very difficult to know exactly how many people are engaged in this activity.

The different types of street trade constitute part of the informal sector because they depend on the traders' own resources, are not established as clearly identifiable businesses and do not declare regular accounts to the authorities (Instituto Nacional de Geografía y Estadística, 2013). Additionally they lack social benefits (e.g. health insurance, annual leave), written employment contracts and receive help from relatives in unpaid work.

In the first quarter of 2016 the number of those in employment in Mexico was nearly 51 million people (45 per cent of the total population). In Mexico City 4 million and 140,000 people (47 per cent of the total population) were registered as in employment (Instituto Nacional de Geografía y Estadística, 2016). Of the total working population in Mexico, 58 per cent work in conditions of informality while in Mexico City the figure is 50 per cent.

To measure the activities of street traders on the public street, the state uses two distinct typologies (Instituto Nacional de Geografía y Estadística, 2011): a) business owners and employers or the self-employed traders in permanent locations in markets, squares, shopping malls or semi-permanent locations in *tianguis* or mobile markets and b) street hawkers without permanent location.

In the first quarter of 2016 there were around 9 million traders in Mexico, of which nearly 7 million are traders in permanent or even semi-permanent locations; the rest are street vendors. In Mexico City, just over 800,000 traders were registered, 77 per cent relate to those with a semi-permanent location, the others are hawkers without permanent location. The significant problem with these data is that they do not identify how many traders are working in the *tianguis* and as we will see below the data, at least in Mexico City, does not match the official reports.

Street trade has different realities, each with specific problems and demands, but all are subject to the regulations for markets which date back to 1951. Since this date, specific rules have been modified in order to create more public space for citizens and eliminate the 'negative effects' of street trade identified by formal entrepreneurs such as: unfair competition, poor quality of products, piracy and tax evasion. Other citizens mention the damage street trade has for the image of the city, problems of vehicular and pedestrian mobility, sanitary problems, corruption and political clientelism which preside over street traders as well as their precarious working conditions.

In addition to the street hawkers, the Mexican government recognises two other types of mechanism for the supply of goods to the working class. The first does not take place on the public street but in shops, public markets and Mexico City's wholesale market, the Central de Abastos (see Delgadillo, this book, for an analysis of covered and municipal markets in Mexico City). The second refers to all the spaces for trade on public streets which are located mainly in marginal areas (Dirección de Abasto, Comercio y Distribución, 2006) and include:

- *'Concentrations' of small traders*: here entrepreneurs conduct their business on semi-permanent stalls, always occupying the same location on land which is either private property owned by the Mexico City government, or is public (Dirección de Abasto, Comercio y Distribución, 2006).
- *'Wheeled' markets*: these are authorised itinerant markets which are installed in different parts of the city in different locations each day; these markets sell food and household products (Dirección de Abasto, Comercio y Distribución, 2006).
- *Tianguis*: these markets mainly offer essential everyday items and are installed one day a week in semi-permanent positions on the public street with a permit that is granted to the representative, the leader, of an organisation (Dirección de Abasto, Comercio y Distribución, 2004). In 2014, there were 1,303 *tianguis* and 171,820 people working in them in Mexico City (SEDECO, 2014).

Tianguis in Mexico City: Disputes, negotiations and rights

Street markets are considered part of the culture and traditions of pre-Colombian Mexico (Martínez, 1985), but most importantly together with the 'concentrations', public markets and 'wheeled' markets they are the means of supply, distribution and marketing of essential everyday items in the outlying areas of the city.

The *tianguis* take place any day of the week in any street or neighbourhood of the city in semi-fixed positions, with metal frame structures roofed with plastic sheeting of different colours which identify their membership of a particular organisation. In these markets vegetables, fruit, meat, fish, poultry, prepared food, clothing and footwear are sold; indeed, any type of product can be obtained.

Generally these markets are set up as not-for-profit organisations that must be registered with the tax authorities, have a company name by which it can be identified, must have members (which in this case are workers of the *tianguis*), hold meetings and have a board. The not-for-profit organisations that represent the traders take care of the permissions to occupy the street and must comply with the legal framework that governs hours of operation, the individual pitches and matters related to safety. This final aspect refers to two areas: 1) Public safety as provided by the state through policing to ensure against theft and violence and 2) civil protection, also the responsibility of the state, to ensure that citizens' activities are carried out under recognised procedures to deal with risks such as using gas, the existence of emergency exits and that the products are safe.

The legal framework of the tianguis

The legal framework that has regulated the *tianguis* since 2004 (Dirección de Abasto, Comercio y Distribución, 2004) has been modified several times in order to ascertain the number and characteristics of the workers and their organisations. In addition it is very difficult to establish the exact number of those working in the *tianguis*. For some years now the government of Mexico City has not granted permission for the creation of new *tianguis* or allowed for an increase in the number of sales pitches, but in reality the existing *tianguis* are growing, with leaders granting more permits and more jobs, and are occupying more of the public space. Given that this is illegal, the leaders of the organisations do not report this growth to the authorities.

For example, in 2010 the recorded number of *tianguis* was 1,420 with 90,000 workers (Asamblea Legislativa del Distrito Federal, 2010). This figure becomes questionable when in an interview that year an official of the municipality of Iztapalapa, one of the poorest in Mexico City, stated that there were 323 *tianguis* and 73,435 traders in that municipality alone and of those only 100 had the necessary permission while the rest were operating illegally although they were tolerated by the authorities. In 2014 the Ministry of Economic Development recorded 1,303 *tianguis* with 171,820 people working in them.

To set up this type of market, three groups of actors are necessary: the first is an organisation that has a charter, statutes and regulations, as well as a register of traders giving their name and the type of product they sell; the second are the traders who besides belonging to an organisation are granted permission to trade by the municipality; the third are the neighbours, who must state in writing their acceptance of the installation of the *tianguis*.

The *tianguis* must comply with basic principles of coexistence, security, hygiene and health and safety. The authorities of each municipality must authorise the permits both for the organisation and the traders, and it must designate the areas where the *tianguis* can take place, distribute the sales pitches, regulate their operation, validate the collection of fees for the use of the street and impose fines in the event of any breach of the regulations (Dirección de Abasto, Comercio y Distribución, 2004).

The tax that the government of Mexico City charges the traders for the use of the street according to the Tax Code of 2015 is 830 Mexican pesos per day (US $0.60) for a semi-fixed position of 1.8 × 1.2 metres. However, in Mexico more than half of the traders that are engaged in this work earn on average 141 dollars a month (CONASAMI, n.d.) and although there is a legally-established quota, it is the leaders of the organisations who define them and charge the minimum of at least the equivalent of one or two dollars a month.

The leaders of the organisations do not necessarily work in the *tianguis* and usually those who carry out the daily business of the markets are the 'delegates' whose main functions are collecting fees for the use of space or 'square', regulating the distribution of individual pitches and maintaining order and solving day-to-day problems.

In 2010 there was a new attempt by the legislative body of the City of Mexico to define clear rules to formalise and 'rescue' (in the government terminology) systems of traditional retail channels such as the public markets and *tianguis*, in order to regulate and organise those involved in the street trade (Asamblea Legislativa del Distrito Federal, 2010). To date, though, this initiative has yet to materialise (Gómez, 2011).

This 'rescue' seeks to ensure that traders are not subject to the control of the leaders and that the negotiation to work in the street is between the traders and the government. It is also intended to prohibit the installation of new markets or a rise in the number of traders, the space that they occupy or the type of products that they sell (i.e. to prevent piracy). The positive side of this initiative is that the traders are not subject to the control of the leaders who monopolise relations with the government and decide, often arbitrarily, on who works and how much they pay for the rent of their pitch. The absence of leaders contributes to the reduction of political clientelism, however historically the leaders have had the ability to defend the stability and permanence of the *tianguis* in the public space.

Disputes and negotiations over public space: The daily life of the tianguis

The street trader's experience of occupying the public space reflects a way of exercising the right to work, although working on the street is precarious, unstable and brings in a low income. However, the experience of trading in the street generates a certain belonging to the public space and a sense of ownership over the sales pitch where traders work every day for many years. For many, the *tianguis* are the only economic activity that they have been involved in, the only thing they know how to do and the idea that 'the street belongs to those that work it' is often spoken of by the traders.

In addition, to be accepted into and to remain working in the *tianguis* demands the acquisition of the necessary knowledge. An example of this is an understanding that it is not the government officials who distribute the spaces but the leaders of not-for-profit associations who, over the years, have established a kind of right and ownership over part of the street. It is the leaders and not the authorities who decide who can operate and under what conditions.

In this section analysis is undertaken in the ways in which traders use, appropriate and contest public space to carry out their economic activity and occupation. This is based on the experiences of some traders in *tianguis* in Mexico City, mainly in the municipality of Iztapalapa, through fieldwork, including participant observation, and interviews.

Interviews with the traders of the *tianguis* reveal the importance of the leaders and of the organisation in the distribution of space, but also for stability in their trade. Some traders with more than 25 years' experience distinguish between *before* and *now* in the entry process of working in street markets: Firstly, before the establishment of traders' organisations and the emergence of leaders, trade in the *tianguis* was extremely unstable due to the constant conflict with the authorities. Besides this, most traders were not set up as businesses, did not have permissions and neither had a fixed or semi-fixed trading pitch and were more like ambulant traders. Secondly, with the shaping of the organisation, the leader becomes a central figure.

Two traders of the *tianguis* explain these two stages:

> So when we started working here, the authorities came and took away the goods, before everyone fought for themselves. If they took my goods from me the person next to me did not care. But it was in 1995, more or less, when there were already quite a few of us, a workmate arrived to sell tacos and we decided to form an organisation.
> (Rosalío, 28 years old, seller of fried potato chips, 06/04/2006)

> Before they took away the stalls and they took everything away from us and they carried off the merchandise, they never returned it, but for some years now agreements were made, over the years things have calmed down, we had many problems with the authorities, but thanks to the support of Mrs Gloria [founder and leader of an organisation in Iztapa-lapa] and all the traders who are united with her, we have moved forward.
> (Paola, 38 years old, seller of fried potato chips, 10/03/2006)

In most organisations the basic function of the leader is to maintain the stability of employment in the public space. To do so the leader acts as the legal representative of the organisation, attends the monthly workshops summoned by the government authorities, must maintain updated information on the number of traders and collect payment every day for the space occupied by the traders. When organisations are large or have *tianguis* in various parts of the city on the same days of the week, leaders designate people or delegates, usually traders from the same organisation, to help in the collection of fees or in the resolution of everyday problems.

The following points explain key aspects of how the *tianguis* work:

- An organisation of traders may have several locations in different neighbourhoods in the city on different days of the week. This can include the sharing of the same space with other organisations as long as they do not coincide on the same day.
- A trader may be working in different locations of the same organisation one, several or all days of the week. At the same time they can belong to several organisations, depending on the *tianguis* and the spatial locations where they have permission and payment of the fee for the pitch.
- An organisation of traders does not have a determined number of members, but it should be at least 100 people as set out in the 1951 regulations of

markets, and can have more than 10,000 people. This applies both in Mexico City and in other states of the metropolitan area.

- Membership of an organisation is obligatory by the simple fact that it is the organisation that controls the space. The trader can decide to leave whenever they want but this results in the loss of their pitch (although in many cases it is possible to sell the space on).
- The role of leader is important in the sale of pitches on public space between the traders themselves. While legally no-one can own a pitch on the street, within the organisations, the space has a very high economic value.

Onésimo, a trader from Iztapalapa, has become a member of an organisation that operates several days a week through the purchase of a space from a woman who has left her stall. He explains that in some *tianguis*, although places are not bought from the leader, it is nevertheless the leader who must agree with the sale, because as the agent they informally monopolise the use of space and also make a profit from it: 'Here, for example, I am owner, the lady who was here before me sold it to me, I asked the price and she asked me for 2,000 pesos per metre and between a neighbour and me we bought 4 metres' (Onésimo, 60 years old, seller of belts). After this description, he states:

> It is not that the leader has to be asked permission to sell but they have to be notified because they have to be given a part of the payment. I paid for two metres and the woman gave the leader 300 pesos. They always get something. Then that I am the owner is only relative because we are on the public street but I had to buy the place.
>
> (Interview, 10/11/2011)

On a day-to-day basis the leader or the delegate[3] is responsible for collecting the rent from the pitches occupied by each trader, ensuring that the set-up and removal times of the stalls are adhered to and that all traders comply with the health and safety regulations. The *tianguis* should have their own generator for electricity as they cannot use the public network, they should monitor the origin of the goods and, in theory, they should not allow the sale of pirate goods or alcohol and must ensure the safety of customers and traders.[4] However, *tianguis* rarely have a generator and 'tap into' existing electricity supplies or ask permission from a neighbour who allows them to connect to their homes. In any case, the delegate ensures that there is an agreement between residents and the traders in order to keep good relations with the community, who are ultimately the main customers of the *tianguis*.

Other important tasks are the resolution of disputes between traders or with the customers and to meet with the authorities when there are inspections. In practice this means accompanying the inspectors, helping to resolve potential fines with the traders and at the end of the inspection inviting them out to eat – which results in a very lax enforcement of the existing regulations.

As with any civil association, there is an obligation to convene meetings where members have the right to participate. However, the assemblies rarely take place, in fact, some respondents commented that when they are held they have never been open to all – only those closest to the leader attend and amongst those are the delegates and some traders with strong connections to the leader. When on rare occasions assemblies are open to all, what is discussed are the everyday problems of the organisation.

Another trader says that they have limited experience of assemblies:

> They are not interested in having them often, or if they do they are for those in the know because they do not advertise them. When a delegate who was the son of the founder leader was appointed a meeting was called but only those closest to the leader went, they passed the motion and he became elected. They have never invited me to a meeting, there is a lot of secrecy, I imagine that if they have an open meeting to which everyone is invited their situation will become more difficult because everyone will begin to talk to one another and concerns will begin to be raised and they want to keep absolute control.
>
> (Ignacio, 46 years old, clothes seller, 11/02/2009)

The *tianguis'* organisations are characterised mostly by a vertical structure in which authority rests with the leader as legal representative and is accepted by the members, because the leader monopolises the distribution of space and establishes relations with the government authorities. The delegate has the authority that the leader grants them.

Leaders operate as the experts of the organisations because they know how to maintain the cohesion of the group, thanks to the skills they have acquired through daily practice and where leadership is often inherited from parents. This capital guarantees control over trade on public streets and the possibilities of negotiation with the government.

In practice the vertical structure works well for the different agents. Members achieve some stability in the occupation of public space, while the leader maintains ties with the political sphere. For their part leaders gain economic and political capital that guarantees the existence of the organisation.

The organisational form of the *tianguis* demonstrates how power relations are established between actors with widely divergent rationales but who coincide on the need to maintain or increase certain forms of capital. The trader leaders and government officials are able to use their political and economic capital to effectively buy, and hence to privatise, public space in Mexico City, although the law does not permit it.

It has been suggested that traders on public streets are strategic actors who learn to adapt to the rules of the game and even resist conflicts with the authorities. However, their scope of action is restricted by economic necessity, which in most cases brings them out to work on the streets with whatever consequences that this has on their working conditions. Additionally, the occupation of public space does not seem to have sufficient political support by those who should regulate it. In the end control over the occupation of the streets by a few, with complicit government authorities, is a very profitable business.

Therefore, despite the fact that street trading allows coexistence and an availability of products, it is necessary to highlight the precarious nature of this occupation and the urgent need to discuss how to make trade on public streets a dignified activity that is part of multiple ways of inhabiting the city. As one trader states:

> It is not because we like working on the street, not because we want to be superior to someone, but because the same need brings us out here [...] those who have other economic possibilities would not sell things on the street, they would go after a stable job; the most viable option for those of us who don't is the street.
>
> (Ricardo, 32 years old, seller of pre-prepared pizza, 05/04/2006)

Conclusions

This research shows the importance and social and academic richness of an occupation that for many people, not only in Mexico City but in many other cities in the world, is part of everyday life. This way of living is part of our 'everyday' and is symbolic of how public space can be appropriated and inhabited.

Commerce in public streets, particularly in the *tianguis*, shows that these streets are not just places where people trade and satisfy their needs, but also spaces for encounter, of coexistence and of political struggle. They are spaces that for many yield financial resources which allow their survival and for some, like the leaders of the organisations, offer possibilities to join the political arena with the construction of political and social actors who use and monopolise the public space at the expense of all citizens.

Another key aspect is the relationship of the legal framework governing the *tianguis* which defines the rules on how to occupy public space and is the link between citizens and the state. The way in which this framework is applied and negotiated demonstrates the absence of adequate laws for the current conditions of street trading, as a source of work and as an activity that uses the space for private purposes. The inefficient legal framework along with a tolerance of its weak implementation contribute to the power of organisations and their leaders and to the discretion practiced by officials and leaders to occupy the public space of Mexico City.

The formation of organisations starts from the basis that it is only possible to achieve collective goals in relations of power, i.e. only in the recognition, although not necessarily the understanding, of the interdependence of the often divergent needs and interests. The field of the organisations that contest public space is a structure of forces characterised by conflict and negotiation between the agents involved. Agents seek to adapt to the structural conditions with the types of capital that each one has; in this case, street trader leaders play with their knowledge of the potential for negotiation and conflict with state institutions.

The various definitions and expressions of trade that take place in the streets invite a reflection on the multiple uses of public space and the power relations that are woven into the struggle for the monopoly of its use and distribution. Taking into account previous research one of the aspects that hinders the understanding of this problem is precisely the complexity of social networks that are involved in trade on the public street.

In these networks, links with agents that make up the political arena, together with the existence of informality and illegality of what is traded on the street, generate a type of 'veil of ignorance' on the part of the officials, leaders of organisations and their associates. This equates to not knowing how many people are involved, who they are and how they occupy the public space and has been useful for the Mexican political system. Through a complex network of complicities, discretionary uses of the law and political clientelism with government authorities, including that of Mexico City and its various municipalities, there is an opportunity for the negotiation of political and economic power between leaders of the organisations and those in power, because the traders in public space can generally be converted into votes at election time. This 'ignorance' has prevented the development of public policies and actions relevant to regulation and even improvement of working conditions for those making a living from trading on the city's streets, and has left unresolved the possibility of guaranteeing basic rights for citizens, to work, to mobility and the right to the city.

Notes

1 Original in Spanish, translated by Neil Turnbull.
2 This has precedence dating back to the 1950s and Ernesto P. Uruchurtu's term as Mayor of the Federal District (Mexico City) (1952–1966) (Cross, 1998).
3 Delegates are appointed by the leaders, who are generally the traders from the same *tianguis*. They are often children, friends or cronies of the leader.
4 Some traders indicated that there was a relatively new problem: groups related to organised crime extorting the payment of fees from leaders of organisations under the guise of 'security' that these groups say they provide. In fact these groups of organised criminals charge the money to ensure they do not 'disturb' traders and customers.

References

Asamblea Legislativa del Distrito Federal (2010). *Mesa de trabajo de la Comisión de Abasto y Distribución de Alimentos*. Available from: http://www.aldf.gob.mx/archivo-312054550182d3e999cee61fb34fc18e.pdf

Bourdieu, P. (1999). Efectos de lugar. In Bourdieu. P. (Ed.), *La miseria del mundo*. Argentina: Fondo de Cultura Económica, pp. 119–124.

Bourdieu, P. and Wacquant, L. (2005). *Una invitación a la sociología reflexiva*. Argentina: Siglo XXI Editores.

Bromley, R. (2000). Street vending and public policy: A global review. *International Journal of Sociology and Social Policy*, 20(1–2), 1–28.

CONASAMI (n.d.). Comisión Nacional de los Salarios Mínimos. Available from: https://www.gob.mx/conasami

Cross, J.C. and Morales, A. (2007). Introduction: Locating street markets in the modern/postmodern world. In: Cross, J.C. and Morales, A. (Eds.), *Street Entrepreneurs. People, Place and Politics in Local and Global Perspective*. Abingdon: Routledge, pp. 1–21.

Cross, J.C. (1998). *Informal Politics. Street Vendors and the State in Mexico City*. Stanford, CA: Stanford University Press.

Dirección de Abasto, Comercio y Distribución (2004). Proyecto de Normas para la Operación de Tianguis en el D.F. Available from: http://www.dgacd.df.gob.mx/normatividad/proyectos/normas_tianguis.html

Dirección de Abasto, Comercio y Distribución (2006). Directorio de Tianguis. Available from: http://www.dgacd.df.gob.mx/comercial/canales/tian.pdf

Gómez, L. (2011). Busca Sedeco modernizar sistemas de abasto en el DF. *La Jornada*, 24 February. Available from: http://www.jornada.unam.mx/ultimas/2011/02/24/busca-sedeco-modernizar-sistemas-de-abasto-en-el-df/

Hiernaux, D. (2013). Tensiones socavadas y conflictos abiertos en los centros históricos: imaginarios en conflicto sobre la plaza de Santo Domingo, Ciudad de México. In: Ramírez, P. (coord.), *Las disputas por la ciudad. Espacio social y espacio público en contextos urbanos de Latinoamérica y Europa*. México: Miguel Ángel Porrúa, pp. 177–198.

Instituto Nacional de Estadística y Geografía (2011b). Sistema Nacional de Clasificación de Ocupaciones, SINCO. México: Instituto Nacional de Estadística y Geografía. Available from: http://snieg.mx/contenidos/espanol/normatividad/normastecnicas/SINCO_2011.pdf

Instituto Nacional de Estadística y Geografía (2013). Informalidad laboral. Presentación técnica. México: Instituto Nacional de Estadística y Geografía. Available from: http://www.inegi.org.mx/est/contenidos/proyectos/encuestas/hogares/regulares/enoe/doc/informalidad_final.zip

Instituto Nacional de Estadística y Geografía (2016). Encuesta Nacional de Ocupación y Empleo. Consulta interactiva de datos. México: Instituto Nacional de Estadística y Geografía. Available from: http://www.inegi.org.mx/Sistemas/Olap/Proyectos/bd/encuestas/hogares/enoe/2010_PE_ED15/po.asp?s=est&proy=enoe_pe_ed15_po&p=enoe_pe_ed15

Maloney, W.F. (2004). Informality revisited. *World Development*, 32(7), 1159–1178.

Martínez, A. (1985). De la metáfora al mito: la visión de las crónicas sobre el tianguis prehispánico. *Historia Mexicana*, 34(4), 685–700.

Morales, A. and Kettles, G. (2009). Zoning for public markets and street vendors. *Zoning Practice. Practice Public Markets*, issue 2, 2–7. Chicago, IL: American Planning Association. Available from: http://urpl.wisc.edu/sites/urpl.wisc.edu/files/people/morales/ZPfeb09.pdf

Ramírez, P. (2013). El resurgimiento de los espacios públicos en la Ciudad de México. Diferencias y conflictos por el derecho al lugar. In: Ramírez, P. (coord.), *Las disputas por la ciudad. Espacio social y espacio público en contextos urbanos de Latinoamérica y Europa*. México: Miguel Ángel Porrúa, pp. 287–314.

Secretaría de DesarrolloEconómico de la Ciudad de México (SEDECO)2014. Reporte Económico 2014. Available from: http://reporteeconomico.sedecodf.gob.mx/index.php/site/main/192

Sethuraman, S.V. (1998). Gender, informality and poverty: A global review. Geneva: World Bank and WIEGO. Available from: http://www.eif.gov.cy/mlsi/dl/genderequa lity.nsf/0/12D2A22FAC60DA74C22579A6002D950A/$file/gender_informality_and_poverty%20(2).pdf

6 Gourmet markets as a commercial gentrification model

The cases of Mexico City and Madrid[1]

Luis Alberto Salinas Arreortua and
Luz de Lourdes Cordero Gómez del Campo

Introduction

This chapter argues that some of the urban transformations undergone by Madrid and Mexico City are related to the substitution of a wealthier sector of the population for another that has less purchasing power, the result being the reconfiguration of space through investment and the transformation of consumption practices. These modifications occur within an urban neoliberal context in which cities have to create the proper conditions to attract private investors, thus generating a constant urban competition. While studies on this gentrification process are mostly focused on the relationship between the transformations undergone by residential spaces and the origins of such a concept, there has been little research on commercial gentrification (though see for example González and Waley, 2013).

Commercial gentrification can be understood as a process in which commercial activity is transformed in order to meet the demands of higher-income groups; such a process involves the displacement of some traders and/or products (Salinas, 2016). In other words, the traditional offer of markets is replaced by the new consumption demands of what Peterson and Kern (1996) call cultural omnivores. These transformations are reflected in the emergence of shopping centres and malls and the restructuring of spaces that focus on traditional commercial activity processes which can lead to the displacement of businesses, traders and/or products in an attempt to create new options to satisfy the demands of higher-income groups.

Markets are part of these traditional commercial spaces that are experiencing transformations to catch up with new the consumption demands described above. Studies that have followed the transformation of markets since the 1990s have been primarily focused on the emergence of supermarkets and on the impact of this process on the decline of traditional markets, which have to compete with these new businesses. There are few studies on the transformation of markets from a commercial gentrification perspective; however, research conducted by González and Waley (2013), who focus on Kirkgate Market in Leeds, England, and Boldrini and Malizia (2014), who focus on the Abasto and Norte markets in Greater San Miguel de Tucuman, Argentina, stand out, analysing as they do the development and implementation of a market model targeted at higher-income groups, especially in the case of *gourmet* markets. Gourmet markets are commercial spaces targeted at casual visitors and tourists rather than local consumers; these markets are also intended to attract a segment of the richest population, which is willing to pay a premium price for having a new 'experience'. These spaces are designed to meet leisure, recreation and selective consumption needs through the adoption of new culinary trends that value healthy, organic and traditional craft production and the provision of 'unique', 'authentic' and 'exotic' foods whose labels are intended to evoke interesting and remote places.

The purpose of this chapter is to broaden and stimulate the debate on the emergence of gourmet markets around the globe as the result of the success of markets such as Mercado de La Boquería (Barcelona), Mercado de San Miguel (Madrid) and Borough Market (London). By using commercial gentrification as an explanatory framework, it is possible to address the transformation of these markets from a space-based class re-structuring perspective.

This chapter is divided into four sections. The first section discusses how the reproduction of gourmet markets can be considered as part of a class process if such a model is intended to attract higher-income and better-educated groups while ignoring most of the population. The second segment of this chapter proposes that the reproduction of gourmet markets is part of an urban competitiveness strategy focused on the attraction of investors and the generation of spaces targeted at residents and tourists with high purchasing power. The third section discusses the specific characteristics of the reproduction of the 'gourmet' model by analysing the cases of Mexico City and Madrid; this analysis is further developed in the fourth section.

This chapter is based on an exhaustive revision of previous research on gourmet markets and new consumption practices, with an emphasis on cultural omnivorousness. Following this, information released by different public entities such as the Department of Economic Development in Mexico City (through its Protection and Promotion Policy for Public Markets, 2013–2018) and the City Council of Madrid through its Innovation Plan 2003–2011 (Madrid City Council, n.d.) was used to study the reproduction of a gourmet market model in Mexico City and Madrid. This exercise also involved the compilation of media reports, particularly from written and electronic sources. In addition, all information collected was complemented with interviews with different key actors, especially with traders from public markets.

Gourmet Markets: Are they part of a class-based process?

The success achieved by gourmet markets such as Mercado de San Miguel (Madrid) and Borough Market (London) has provoked an interest to replicate them in different markets around the globe. There are different cities that offer gourmet, craft or organic markets as tourist attractions. This is the case in San Francisco, USA, where one of its main markets, Ferry Building

Marketplace, has forged a strong alliance with local producers of artisan foods (Ferry Building Marketplace, 2016); in Valencia, Spain, where Mercat Central does not only offer fresh local products, but also represents a cultural and tourist attraction for foreign visitors and local residents alike (Mercat Central de Valencia, 2012); and in Genoa, Italy, where the Mercato Orientale is promoted as a shopping centre that is visited by local users and tourists from over 75 different countries (Mercato Orientale di Genova, 2015). This trend is part of a global process in which highly-educated consumers with high purchasing power have expanded their cultural and culinary experiences in order to consume 'authentic' and 'exotic' products labelled as 'natural', 'organic', 'simple', 'traditional', 'artisan-made' or 'locally produced'; these users want to live a gourmet experience in their local communities and also include this activity in their preferred holiday destinations.

These new preferences of consumers have been studied by authors such as Peterson and Kern (1996), who refer to such a phenomenon as *omnivorousness*. This concept describes the existence of 'a segment of the population from western countries that implements and makes choices from a wider option of cultural forms than those previously available, thus reflecting new values associated with tolerance and weakening the concept of snobbism' (Warde et al., 2007, p. 143). For cultural omnivores, current preferences and traditional values, such as those associated with the French haute cuisine, are no longer the sole references of this trend. This has led to an increase in the supply of products produced in different cultural contexts in order to meet the needs of an emergent market, the result being the proliferation of gourmet markets.

However, these new preferences may not be strictly related to the decline in snobbishness, as is claimed by Warde et al. (2007), but to a modification in the consumption patterns of elite users. According to Hyde (2014), class distinction and the search for status determine the behaviour of cultural omnivores. In this line, Johnston and Baumann (2007) point out that the access to these new cultural preferences depends on the economic and cultural capital of individuals; this means that the new culinary offer is exclusively targeted at the upper-middle and upper classes, generating the same exclusionary patterns that prevailed when French *haute cuisine* was the only existing gourmet trend.

When visiting gourmet markets, consumers are able to share that 'authentic' and 'exotic' experience with individuals in other parts of the globe, such as Madrid or London, who are also enjoying these new preferences and culinary habits that connect them with a social class that understands the cultural importance of certain foods and products. These consumers, through social networking channels such as Instagram or Snapchat and travel or lifestyle media, actively promote the exotic nature of these products by labelling them as interesting or unusual and relating them to foreign places and customs or adding adjectives like extravagant, scarce, hard to find and rare. All these features are used to sell the items offered at gourmet markets which, in turn, promote this type of consumption that has been labelled as the 'gourmet market model'. Within this context in our own research we are interested in answering the following questions: what lies behind the 'reproduction' of a gourmet market model? Why do the transformations undergone by traditional markets aim at replicating the aforementioned market models? How is this transformation implemented in different contexts? These questions can be addressed by examining the experiences of cities such as Mexico City and Madrid.

Gourmet markets and the competitive advantage of cities

During the second half of the twentieth century, the central areas of the main European cities began to experience a gentrification process. Gaja (2001) refers to that phenomenon as the recovery of historic sites, which can be traced back to the late 1960s. This process involved the 'recovery' of historic and/or symbolic places, such as public markets. For instance, the Mercado de San Miguel (Madrid) is located in the space once occupied by the church of San Miguel de Octoes, which stood there during the eighteenth century until 1790, when it was partially destroyed by a fire and then torn down by order of King Joseph Bonaparte (Anasagasti, 1916). This area, which was then used as a public square, contained a market that offered perishable products. By the end of the nineteenth century, and according to the then-new Tourist Plan, this space was used to build an indoor market made of iron that drew inspiration from markets such as Los Mostenses and La Cebada, both built in 1875. The new Mercado de San Miguel, which was made of iron and glass, was completed in 1916 (Madrid Histórico, 2003). This market operated successfully until the 1980s when it was forced out of business for 10 years as the result of the emergence of supermarkets. In 1999, the market was modernised thanks to resources provided by local traders, the European Union and the Regional government of Madrid; in 2000 it was declared a national heritage monument with the highest protection levels. However, it was not until 2003 that a group of private investors purchased this space and transformed it into a gourmet market. The new establishment reopened in 2009 and became an attraction for tourists and local residents (Mercado de San Miguel, 2013). This market model has been replicated in different cities such as Mexico City and opened its first franchise in the city of Miami in 2015 (La Feria Mercado de San Miguel, 2014).

Today, cities compete with each other to attract new inhabitants and tourists with high purchasing power. They are focused on the development of consumption-related competitive advantages which transform local markets into 'major tourist attractions' (Medina and Álvarez, 2009, p. 195). In this sense, the gourmetisation of the urban offer in restaurants, markets or food trucks is a response to such a phenomenon.

Against this competitive backdrop, Richard Florida (2003) has suggested that cities should attract the *creative class*, which is the driving force for regional economic development. Florida points out that highly-educated people are attracted by inclusionary places that embrace diversity. Since the 'creative class' is composed of highly-educated people, cities should attract individuals holding higher education degrees. An economically vibrant and internationally competitive city promotes values such as diversity, tolerance, inclusion and cultural enlargement. Competitive cities – focused on promoting themselves to attract investors, residents and tourists – are spaces that offer different, varied, creative, inclusionary, authentic and exotic options for capital reproduction. Such a discourse is also found in the reproduction of gourmet market models.

Since gourmet markets are an important component of the commercial and cultural activities to be offered by creative cities, authorities may modify or create new public policies to support the transformation of traditional markets. This promotes an 'an urban imaginary that regards the consumption of gourmet food as an essential component within the creation of an exciting urban experience' (Martin, 2014, p. 1880). However, this discourse – which is based on concepts like creativity, inclusion, tolerance and diversity – is only addressed to certain segments of the population. In her paper 'Food Fight' (2014), Nina Martin describes the contrasting experience of immigrants (mostly Mexican), street vendors and Food Truck chefs, in which the latter were able to modify the restrictive policies governing street vending activities in the city of Chicago, thus achieving business success. It seems that the implementation of inclusionary and tolerance-oriented policies is only intended to favour some segments of the population that receive the support of politicians, provided they offer a wider array of options to cultural omnivores.

The marketing of cities also involves the promotion of gourmet markets, which are transformed in order to attract cultural omnivores – either local residents or tourists – by offering an experience associated with concepts such as diversity and authenticity. This is why successful gourmet market models such as Borough Market (London), Mercado de San Miguel (Madrid) or Mercado de La B/oquería (Barcelona) – regarded as some of the best markets in the world according to magazines and specialised websites such as Travel and Leisure, The Daily Meal, CNN and Travel Channel – are being replicated in different cities around the globe. These new markets are not only focused on selling the same products offered by their original counterparts, but also on replicating the way these goods are displayed and marketed. The gourmet market model has become an export product.

In his discussion of the visual culture observed at Borough Market (London), Coles (2014) suggests that this place aims at creating an experience based on the production of cultural material through its transformation into a visual economy since 'Borough Market has been designed to look and feel "market-y". The different stalls, signage and other visual images generate geographical and imaginary associations with the origins of food and how it should be consumed' (Coles, 2014, p. 520). In the words of this author, Borough Market is a real 'spectacle'. The same approach may be applied to Mercado de San Miguel (Madrid), which has become a tourist attraction that – set within a natural and unique context – offers a particular experience associated with authentic and exotic elements by using local, regional or remote geographical references. This has led to a growing interest in importing and reproducing this model since gourmet markets are capable of increasing the culinary options of competitive cities.

However, the local characteristics of markets and their political and economic contexts affect the reproduction of gourmet market models, which is often only partially achieved. Hence, rather than the full implementation of a model, what we find is that some markets offer gourmet products; in other words, rather than the adoption of a market model we see a trend for the gourmetisation of consumption.

The reproduction of gourmet market models

In recent years, there has been a steep increase in the import/export of urban policies labelled as successful – including gourmet market models – between cities and countries. The replication of urban policies considered successful such as those of Vancouver, New York or Barcelona in different cities of Europe, Asia and Latin America has given rise to the emergence of an industry of import/export of urban policies from city to city.

González (2011) refers to this practice as 'urban policy tourism' in an attempt to draw a parallel with traditional tourism in the sense that policy tourism is also based on networks that host visitors in terms of itinerary planning and the provision of experts to inform and guide them through the city. In this sense, both investors and public officials are able to experience the success stories of these places – which have been disseminated by the media, specialised literature, documents, social networks, pictures and exhibitions – *before* implementing them in their respective cities (González, 2011).

However, it should be noted that there is no automatic association between the formulation and implementation of such policies. In other words, local characteristics (political, economic and social conditions) and the transformations undergone by these policies during import-reproduction processes should be taken into consideration (Peck and Theodore, 2010), since what is successful in one place might not work in another. Such is the case of Mexico City, where private gourmet investors and the local authorities in charge of the administration of public markets in the municipality of Álvaro Obregón travelled to Madrid and expressed interest in reproducing the successful Mercado de San Miguel model without considering the feasibility of its implementation in a completely different economic, political and social environment. In Barcelona, the Institute of Municipal Markets has launched a department of international relations in an attempt to assist the growing number of foreign urban policy-makers and politicians who want to understand the so-called 'Barcelona market model' (Instituto Municipal de Mercados de Barcelona, n.d.).

Ward (2006) points out that imported policies are implemented in real places and require specific strategies in order to succeed. At least this is the result that authorities and private interests hope when they import a 'market model'; however, a given solution may not work in different scenarios. This means that the successful conversion of these initiatives requires more effort than the importation of a strategically modified model that may not ensure the expected outcome. As McCann (2011) suggests, the differential access to resources determines the success of each policy, so there is a need to understand the contexts in which they are reproduced, transferred and adopted.

The application of a 'gourmet market' model through the importation of public policies that have worked in one place does not ensure a positive outcome, since they have not been adapted or developed to operate within a given local context; therefore, rather than being imported just because of their success, the implementation of these models should include the analysis of positive and negative impacts as well as a feasibility study.

Gourmet market models in Mexico City and Madrid

According to data released by the Secretariat of Economic Development, there are 329 public markets containing more than 70,000 businesses in Mexico City (2013). These markets have witnessed the migration of customers to different outlets such as supermarkets. This phenomenon and the lack of investment in infrastructure and the provision of services have led to 'a significant decline in economic activity and visitor numbers' according to the local economic development agency of Mexico City (Secretaria de Desarrollo Económico Gobierno de la Ciudad de México, 2013, p. 4).

Public markets are an emblematic component of the popular economy in Mexico City, generating 200,000 jobs (Protection and Promotion Policy for Public Markets in Mexico City, 2013–2018, p. 4). This is why the government of Mexico City has listed these markets as 'an economic, urban and cultural priority' (ibid, p. 4) and developed protection and promotion policies that include a modest remodelling budget to attract old and new customers.

As some mayors and politicians have expressed, these measures aim to replicate successful foreign models such as Borough Market (London) or Mercado de San Miguel (Madrid), implementing the gourmet market concept. According to Leonel Luna, Mayor of the municipality of Álvaro Obregón, the redevelopment of public markets is intended to offer refined products and modernise these spaces, not only through physical renovation, but also through the installation of ATMs and extended service hours. This discourse is based on the premise that there is a need to address the deterioration and decline in the economic activity of these markets as the result of the emergence of supermarkets.

With a budget of USD 500,000–900,000, the City Council of Álvaro Obregón – in conjunction with two private universities (Iberoamericana and Monterrey Tech) – will conduct research on the situation of its 14 markets, which will be then redeveloped accordingly (Montes, 2014). The City Council of Álvaro Obregón is intended to reproduce the gourmet model through the renovation of the Mercado Melchor Muzquiz (Montes, 2014). According to City Council officials, the intervention of Mercado Melchor Muzquiz – commonly known as Mercado de San Ángel and located in an exclusive and tourist area – will serve as a pilot project for the implementation of the gourmet model. Built in 1958, this space was recently renovated as part of the Program for the Improvement of Public Markets. In 2014, a City County official said that Mercado Melchor Muzquiz – which offers a type of architecture and products that may fall within the gourmet category – 'may be transformed into an historic, tourist and gourmet centre while yet retaining its nature in a space similar to that of Mercado de San Miguel, Spain' (Montes, 2014). Despite these intentions, it is not clear whether or not Mercado de San Ángel will become a gourmet market. What is apparent, though, is the intention of this market to offer 'different' and 'exotic' products to attract higher-income groups.

In the case of Mexico City, gourmet markets emerged in 2014 after the opening of Mercado Roma, which is located in the neighbourhood of the same name. However, unlike Borough Market – which has been operating for more than 200 years– the space used by Mercado Roma had to be modified, since it was not initially designed to house a market; it is located in a former dance bar that stood there for more than four decades.

Shortly after the opening of Mercado Roma, Mexico City witnessed the inauguration of Mercado del Carmen, the second 'gourmet market'. Located in the neighbourhood of San Ángel, this market is housed in a colonial-era building that was formerly used as an art gallery. Recently, the neighbourhood Colonia Polanco saw the opening of La Morera, a gastronomical centre located at the intersection of Palmas and Mazaryk Streets. Despite not being labelled as a market, this centre is also intended to replicate the gourmet market model. As of today, the new gourmet markets inaugurated in Mexico City are located in spaces that were not designed to house such an activity: Milán 44, Casa Quimera, Mercado Moliere and Mercado Gourmet Samara.

These examples show that, rather than observing the transformation of traditional markets into gourmet markets, there is an emergence of consumption spaces that are being labelled as gourmet markets in order to attract middle and higher income groups. The aspiration of becoming a gourmet market is based on the replication of consumption spaces regarded as successful by the business sector, such as the case of Mercado de San Miguel (Madrid). This is explained by one of the founders of La Morera, 'the idea started during a visit to Madrid. The old Mercado de San Miguel was unrecognisable because of its modernised structure and wide array of culinary options' (Forbes Staff, 2015). According to its founders, La Morera was created by drawing inspiration from Mercado de San Miguel.

There are two trends associated with the generation of the so-called gourmet consumption spaces in Mexico City. On the one hand there is the proposal of entrepreneurs, who focus on the 'staging' of markets. This leads to the generation of the 'gourmet market' concept, which is intended to attract consumers with sophisticated tastes and high purchasing power. These spaces are promoted as markets but they are not regarded as such. Despite lacking the history and the social and economic networks of their traditional counterparts, they pretend to have all these features and operate in spaces that were not designed to house markets; hence the word 'staging'. On the other hand, local governments are focused on transforming traditional markets into gourmet markets; however, the intervention in Mercado de San Ángel has failed to achieve that objective.

Likewise, it is also possible to identify two trends in Spain. On the one hand, local governments aim to renovate and transform municipal markets in order to address the deterioration and decline in economic activity generated by the emergence of supermarkets and new consumption habits and, on the other, there are the conditions for the generation of recreational spaces through the concept of the gourmet market and situations that favour the emergence of staged markets.

Besides addressing their deterioration and decline, these strategies are also intended to transform traditional markets into gourmet-based markets. Inspired by the experience of Mercado San Antón and Mercado de San Miguel (Madrid), Iñigo de la Serna – Mayor of Santander (a city in northern Spain) – is planning to turn the city's Mercado de Puertochico into a recreational and tourist area, which is also intended to become a 'landmark and an energising force for the neighbourhood' (Lemaur, 2013). The City Council of Santander has allocated 310,000 Euros for the renovation of the public square that surrounds the market; this project also includes the restoration of the inner section of Mercado de Puertochico, which is expected to be completed by 2018 (El Diario, 2016).

In the case of Madrid, municipal markets are expected to meet the new consumption demands through the implementation of different public programmes. In this sense, the Innovation Plan 2003 suggested that municipal markets were showing signs of decline and obsolescence in relation to the new commercial trends. This situation, in which municipal markets were regarded as obsolete, took place before the implementation of further transformations.

The transformation of Mercado San Antón, located in central Madrid, was preceded by the imposition of a discourse based on the deterioration of this market (Salinas, 2016; see also García et al, this book). The new Mercado San Antón included the construction of a tasting area that was not regulated by the then-current legislation; such an initiative was made legal once the inauguration process was over. These amendments show how commercial transformations respond to the emergence of new consumption practices; tasting areas for tourists, spaces for the sale of gourmet products, bars and cafeterias are built in commercial areas targeted to individuals with high cultural and economic capital who prioritise the use of these sites and services over the acquisition of basic food. This is an example of the reproduction of a leisure area that offers 'authentic' options within a marketing context. Meanwhile, the Mercado Los Mostenses, also in a central location, is intended to be transformed, according to the president of Los Mostenses Traders' Association, into 'a market of the twenty-first century' by shedding its traditional, old-fashioned image (Salinas, 2016; but see also García et al., this book, for a detailed analysis of this market).

On the other hand, the staging of gourmet markets, as in the case of Mercado Roma in Mexico City, is also being replicated in Spanish cities. In 2012, Madrid saw the opening of La Isabela, which, labelled as a gourmet market, reproduces the experience of Mercado San Antón. In 2015, Puerta Cinegia Gastronómica was inaugurated in the city of Zaragoza; according to its spokesman, José María Ortiz, this market took inspiration from Mercado San Antón (Madrid) since this is the city that witnessed the emergence of these spaces that combine food, gastronomy and recreation. 'Though it emerged in Madrid, the origins of this trend can be said to be linked to Mercado de La Boquería in Barcelona' (Puerta Cinegia Gastronómica, 2015).

In Madrid, Mercado de San Miguel – the first of its kind in the city – and the consolidation of Mercado San Antón have served as inspiration for the creation of gourmet markets according to two trends: the transformation of municipal markets and the staging of markets. Some examples of the first case are Mercado Barceló – which is composed of a market, a sports centre and a library – and Mercado de Chamartín – which refers to itself as an 'authentic gourmet market' and defines gourmet markets as: 'those that offer select and exclusive products so that the authentic connoisseurs will find them as perfect as their places of origin, where they are raised, cultivated and collected' (Mercado de Chamartín, 2015).

As for staging markets, they are easily reproduced in places such as Mercado de San Idelfonso, which promotes itself in their website with the slogan 'we are not just another market' (Mercado de San Idelfonso, n.d.) and adopting a street vending approach as in the case of London and New York-based markets. This type of market is also used to launch books and commercial products, hold work meetings, courses, exhibitions, projections, filming and photographic sessions, cooking shows, etc. Another case is that of *Platea Madrid*; located in the former Carlos III cinema and defined as a space for gastronomic recreation, this market had to be renovated in order to house this new type of activity.

The transformation of municipal markets and the staging of recreational and consumption spaces characterise the commercial changes that take place in Madrid. These processes are intended to meet the unique and exotic culinary demands of middle- and higher- income users.

Conclusions

The gourmet market model is being replicated as part of a trend known as the gourmetisation of consumption and as part of an urban neoliberal business strategy designed to meet the new culinary preferences of foreign and local cultural omnivores. As a result, these spaces cannot be accessed by large segments of the population since they are not able to afford gourmet products.

In this sense, traditional markets aim at becoming gourmet markets within a context of urban competitiveness. This has allowed cities to develop consumption-related competitive advantages and emerge as major tourist attractions. The gourmetisation of consumption in restaurants, markets or food trucks is a response to such a phenomenon.

In Mexico City, we do not find real evidence from local authorities of the import and reproduction of gourmet market models such as those recognised internationally; there is an intention in their discourses but this is not followed up. However, though far from replicating such a model, it is possible to observe a gourmetisation of consumption habits. Aside from the public authorities, private initiatives have been focused on the staging of urban spaces – which are promoted as markets – in an attempt to meet the gourmet demands of higher-income users. As of today, none of these spaces has been able to replace traditional markets.

Likewise, the Mercado de San Miguel has served as an inspiration for the transformation of other municipal markets such as San Antón. There are also spaces that, as in the case of Mexico City, have been modified to resemble a gourmet market; these spaces are intended for recreation and selective consumption, with the trade of basic food being relegated to a secondary role. These commercial markets are run by higher-income groups and targeted to higher-income groups.

In both cities, imported gourmet models have met with different degrees of success, from failed attempts – such as the Mercado Melchor Muzquiz (Mexico City) – to the staging of markets in places that were not intended to house this type of activity – the Mercado Roma (Mexico City) and the Mercado San Idelfonso (Madrid). However, both cases aim at achieving the gourmetisation of consumption practices targeted to higher-income groups.

Note

1   Original in Spanish translated by Juan Pablo Henríquez Prieto

References

Anasagasti, T. (1916). La Construcción en Madrid. El Mercado de San Miguel. *Revista Construcción Moderna*. Available from: www.mercadosanmiguel.es/blog/almacen/la -construcción-en-madrid-el-mercado-de-san-miguel-por-teodoro-anasagasti-1916/
Boldrini, P. and Malizia, M. (2014). Procesos de Gentrificación y Contragentrificación. Los Mercados de Abasto y del Norte en el Gran Tucumán (Noroeste Argentino). *Revista INVI*, 29, 157–191.
Coles, B.F. (2014). Making the marketplace: A topography of Borough Market, London. *Cultural Geographies*, 21(3), 515–523.

El Diario (2016). La plaza del Mercado de Puertochico pasa de ser 'invisible' a abrirse a la ciudad con un diseño 'atrevido'. *El Diario.es*, 1 March 2016. Available from: http://www.eldiario.es/norte/cantabria/ultima-hora/Mercado-Puertochico-invisible-abrirse-atrevido_0_489951413.html

Ferry Building Marketplace (2016). Main website. Available from: http://www.ferry buildingmarketplace.com

Florida, R. (2003). Cities and creative class. *City and Community*, 2(1), 3–19.

Forbes Staff (2015). La Morera, un centro gastronómico fuera de lo común. *Forbes México*, 25 March 2015. Available from: http://www.forbes.com.mx/la-morera-un-centro-gastronomico-fuera-de-lo-comun/

González, S. (2011). Bilbao and Barcelona 'in motion'. How urban regeneration 'models' travel and mutate in the global flows of policy tourism. *Urban Studies*, 48(7), 1397–1418.

González, S. and Waley, P. (2013). Traditional retail markets: The new gentrification frontier? *Antipode*, 45, 965–983.

Hyde, Z. (2014). Omnivorous gentrification: Restaurant reviews and neighborhood change in downtown eastside of Vancouver. *City & Community*, 13(4), 341–359.

Instituto Municipal de Mercados de Barcelona (n.d.). Modelo Mercado Barcelona. Ayuntamiento de Barcelona. Available from: http://ajuntament.barcelona.cat/merca ts/es/canal/model-mercat-barcelona

Gaja, F. (2001). Intervenciones en los centros históricos de la Comunidad Valenciana. Consejería de Obras Públicas, Urbanismo y Transporte. Dirección General de Arquitectura y Habitación. Universidad Politécnica de Valencia, Departamento de Urbanismo. Available from: http://personales.upv.es/fgaja/publicaciones/cen troshistoricos.pdf

Johnston, J. and Baumann, S. (2007). Democracy versus distinction: A study of omnivorousness in gourmet food writing. *American Journal of Sociology*, 113(1), 165–204.

La Feria Mercado de San Miguel (2013). Website. Available from: www.mercadodesa nmiguel.us

Lemaur, V. (2013). El Mercado de Puertochico podría transformarse en un espacio de ocio y restauración. *El Diario Montañés: El Diario de Cantabria y Santander*, 2 Febrero 2014. Available from: http://www.eldiariomontanes.es/20130202/local/santa nder/mercado-puertochico-podria-transformarse-201302021307.html

Madrid City Council (n.d.). Plan de innovación y transformación de los mercados de Madrid: 2003–2011 ocho años impulsando el comercio [online]. Madrid: Madrid City Council. Available from: http://www.madridemprende.es/images/public/sour ce/plan_mercados.pdf

Madrid Histórico (2003). El Mercado de San Miguel. *Centro de Documentación de Historia de Madrid de la Universidad Autónoma de Madrid y Desarrollo, Asesoría y Formación Informática S.A.* Available from: www.madridhistorico.com

Martin, N. (2014). Food fight! Immigrant street vendors, gourmet food trucks and the differential valuation of creative producers in Chicago. *International Journal of Urban and Regional Research*, 38(5), 1867–1883.

McCann, E. (2011). Urban policy mobilities and global circuits of knowledge: Toward a research agenda. *Annals of the Association of American Geographers*, 101(1), 107–130.

Medina, X. and Álvarez, M. (2009). El lugar por donde pasa la vida … Los mercados y las demandas urbanas contemporáneas: Barcelona y Buenos Aires. *Estudios del hombre*. 24, 183–201.

Mercado de Chamartín (2015). Website. Available from: http://www.mercadodechama rtin.es/

Mercado de San Idelfonso (n.d.). Website. Available from: http://www.mercadodesanil defonso.com/

Mercado de San Miguel (2013). Website. Available from: www.mercadodesanmiguel.es

Mercat Central de Valencia (2012). Website. Available from: https://www.mercadocen tralvalencia.es/

Mercato Orientale di Genova (2015). Website. Available from: http://www.mercatoor ientale.org/

Montes, R. (2014). San Ángel tendrá mercado tipo gourmet, revela delegado. *El Financiero*, 2 March 2014. Available from: http://www.elfinanciero.com.mx/socieda d/san-angel-tendra-mercado-tipo-gourmet-revela-delegado.html

Peck, J. and Theodore, N. (2010). Mobilizing policy: Models, methods, and mutations. *Geoforum*, 41, 169–174.

Peterson, R. and Kern, R. (1996). Changing highbrow taste: From snob to omnivore. *American Sociological Review*, 61, 900–907.

Puerta Cinegia Gastronómica (2015). Website. Available from: http://www.puertacine giagastronomica.es/

Salinas, L. (2016). Transformación de mercados municipales de Madrid. De espacio de consumo a espacio de esparcimiento. *Revista INVI*, 85(31), 179–201.

SEDECO (2013). *Política de Protección y Fomento para los Mercados Públicos de la Ciudad de México (2013–2018)*. Ciudad de México: Secretaria de Desarrollo Económico Gobierno de la Ciudad de México.

Ward, K. (2006). 'Policies in motion', urban management and state restructuring: The trans-local expansion of business improvement districts. *International Journal of Urban and Regional Research*, 30(1), 54–75.

Warde, A.David Wright, D. and Gayo-Cai, M. (2007). Understanding cultural omni-vorousness: Or the myth of the culture omnivore. *Cultural Sociology*, 1(2), 143–164.

7 Neighbourhoods and markets in Madrid

An uneven process of selective transformation[1]

Eva García Pérez, Elvira Mateos Carmona, Vincenzo Maiello and Alejandro Rodríguez Sebastián

Introduction

This chapter addresses the process of transformation experienced by markets in Madrid over the last decade. Such an evolution is framed within an urban context where the global economy and the impacts of the selective action of public policies (sometimes intervening and other times ignoring markets) are regarded as key forces of change. This study focuses on the municipal network of markets of Madrid paying specific attention to three markets. Using empirical methodology, observation and qualitative and quantitative analysis, this chapter presents the transformation of markets in Madrid not as one single process but as an array of possibilities. According to our perspective, these transformations are closely related to the urban context in which markets are located, i.e. their population, economic and social contexts. In recent years, markets have lost their function of being local and neighbourhood-based food suppliers to become specialised spaces and this has fragmented them into different categories that we will explore later: the traditional market, the gourmet, the organic and the service-oriented market. We refer to this variegated process as a *selective* transformation. In order to further illustrate these changes, this research focuses on three paradigmatic cases of markets located in downtown Madrid: Mercado de San Antón (in the Chueca district), Mercado de los Mostenses (in Maravillas) and Mercado de San Fernando (in Lavapiés). The objective of our research is both to highlight the close relationship that exists between urban and commercial dynamics in spaces where gentrification activities converge at the same time and at different levels and to identify the current renovation of markets as a *retail gentrification* phenomenon. The conclusions of this chapter present markets as new *border* spaces that strengthen and allow the development of gentrification dynamics which, ultimately, can lead to the displacement of lower-income groups – traders and consumers – and the coexistence of different lifestyles and social classes.

Theoretical background on commercial gentrification

While studies on gentrification are mainly focused on the analysis of residential transformations, there are some authors that establish a relationship between commercial changes and population and class restructuring processes that take place in cities of the Global North such as San Francisco, Toronto, Portland, Sydney, Paris, Lyon, Amsterdam and Porto. In these studies, the classic approaches used in economic and urban geography meet with consumption sociology (Lemarchand, 2011). In this way, the traditional studies that linked retail activities directly to forms of production and consumption and associated them with urban systems have been broadened by studying the new ways of life and consumption habits of urban citizens. The symbolic aspects of consumption are also gaining importance when it comes to describing the gentrification phenomenon (Lees et al., 2008; Hernández Cordero, 2014). In the case of markets, previous studies analysed them in terms of the consumption of symbolic goods (Crewe and Beaverstock, 1998). This goes in the same direction as other research which described consumption as an *experience* in itself (González and Waley, 2013; Smithers et al., 2008), where the social practice of purchasing food has a symbolic meaning that adds extra value to products.

Most of the current research on commercial gentrification focuses on the study of streets (Fleury, 2003; Lehman-Frisch, 2002; Roth and Grant, 2015) or neighbourhoods (Authier, 1989; Bridge and Dowling, 2001; Hackworth and Rekers, 2005; Omhovère, 2014; Varanda, 2005; Zukin et al., 2009) where trade has a significant presence and there is proof this activity is undergoing a reorganisation process. In many occasions, this commercial restructuration is closely related to urban renovation initiatives (Rankin and McLean, 2015) and in most of these cases there is a decrease in neighbourhood trade activities in central areas (Authier, 1989; Lehman-Frisch, 2002; Varanda, 2005). Our research focuses on the way that markets, despite their architectural particularities and indoor nature, are incorporated into these urban dynamics of commercial gentrification (González and Waley, 2013).

The current literature on commercial gentrification intersects with several different issues. Firstly, there is the role of trade, traders and users as key actors in urban transformation processes (Berroir et al., 2015) as well as the intermediary role of new investors and entrepreneurs (Zukin et al., 2009) in public places where economic activities take place. Likewise, it is worth mentioning the importance of public policies in determining and steering these changes, either by protecting traditional trade, pushing it to become more entrepreneurial or creating business improvement areas or districts (Hackworth and Rekers, 2005) or by treating actors in very different ways, as in the case of migrant street vendors (Martin, 2014). These issues are also related in terms of the control or promotion of gentrification (Viana Cerqueira, 2014).

There are other studies that see the value of neighbourhood trade as an essential resource for the low-income and working class groups (Chabrol,

2011; Omhovère, 2014). Watson (2009) refers to markets as useful and vibrant spaces where social inclusion and the recreation of social relationships turn them into socially-rich and interesting places. According to Spanish authors Guàrdia and Oyón (2007, 2010), trade is an element that builds social relationships. As the result of retail transformations, the reactions of traders and the population may range from survival and adaptation strategies to resistance (Authier, 1989; Sullivan and Shaw, 2011; Varanda, 2005).

Likewise, some studies suggest that trade is an identity sign that defines urban culture (Bridge and Dowling, 2001); in this sense, attention should be given to two aspects: distinction (Shkuda, 2013) and multiculturalism (Semi, 2005). These elements may be involved in a contradictory relationship that refers to values such as authenticity (Van Criekingen and Fleury, 2006) and commodification (Hackworth and Rekers, 2005). The democratisation of luxury and the increase in values triggered by consumption (Michaud, 2015) clearly explain the occurrence of gentrification processes, the use of culture and arts as distinction mechanisms in certain neighbourhoods and the gourmetisation of food trade in different markets around the globe. Lastly, the displacement issue has been directly related to ethnicity and class (González and Waley, 2013; Sullivan and Shaw, 2011; Zukin, 2009).

The above allows us to highlight the main role of trade within the context of cities. In this sense it is worth mentioning structural aspects, since trade has historically been a key element in capital reproduction processes, as pointed out by Harvey (2008) in the case of nineteenth century Paris. Today global changes have significantly influenced the new production and provision models controlled by large oligopolies (Schwentesius and Gómez, 2006), thus critically affecting traditional models of provision and consumption. Both issues take place within an urban neoliberal context (Brenner, Peck and Theodore, 2010) where the international mobility of public policies contributes to the import of models considered successful and closely associated with city branding strategies (González, 2011). If during the 1990s, leisure and consumption landscapes were dominated by malls today, however, the emergence of new aesthetic and spatial dimensions is redefining shopping experiences. New formats (such as boutiques, farmers markets and food fairs) add extra value to the recovery of certain aspects of traditional trade – with a focus on the provision of high-quality products, expertise and warm relationships – within a culturally redefined landscape (Mermet, 2011).

As for the Spanish case, Barcelona – with its successes and failures – is a clear example of market renewal policies (Barcelona City Council, 2015). Within a context marked by unequal geographies, these transformations have given rise to new urban renovation and profitability opportunities derived from commercial rent gaps (Smith, 1987 and 1996).

In the case of Madrid, a commercial overview of the 1990s (Checa and Lora-Tamayo, 1993) reveals the spatial and sectoral structure of local trade activities. The importance acquired by the commercial function of the urban centre since the beginning of the nineteenth century is associated with non-homogeneous spatial patterns, whose distribution coincides with the predominance of food and/or popular trade in the neighbourhoods that we are investigating. Today, these neighbourhoods show evidence of ongoing gentrification activities (García Perez, 2014).

A genealogy of markets and public policies in Madrid

Decline and re-emergence of markets in Madrid

Madrid has a large network of retail markets built in successive stages over the twentieth century. The initial phase, which lasted from the second half of the nineteenth century to the 1920s, was characterised by the emergence of the first municipal policies on the sale of food and the construction of the first cast-iron market buildings; this period was also marked by the presence of different types of markets (indoor, private, public, open-air and granted in concession). This was the starting point of a series of public policies concerning markets that transcended different historical and political moments. The year 1929 saw the elaboration of the first plans for the rationalisation and expansion of markets and then, in 1943, the *Programme for the Construction of New Markets* was launched, with the goal of establishing a network composed of 26 neighbourhood markets, including new and already existing spaces. The following decades witnessed the extension of this network, which grew hand in hand with the city until the 1980s; as of today, this network is composed of 46 municipal markets with a presence in 17 out of the 21 districts of the city. During this period, markets in Madrid operated as non-exclusionary 'cross-class' spaces that ensured the provision of essential goods. Given their price-setting role and the promotion of competition, markets were privileged spaces where lower-middle and lower classes did their grocery shopping. Likewise, low-cost rents enabled the emergence of family-based and neighbourhood-oriented small businesses (Bahamonde Magro and Fernández García, 2008; López de Lucio, 1998, 2007; Montoliú Camps, 1988). As will be discussed below, the current transformation is affecting these traditional forms.

The first signs of a crisis for markets were detected in the early 1990s. Since then, markets have been constantly blamed for their inability to adapt themselves to the new habits of consumers. In the official narrative, the emergence of new tastes and social trends is regarded as the result of natural and spontaneous decisions and markets are blamed for their mismanagement and inability to modernise themselves due to the 'obsolescence of structures in relation to modern commercial distribution trends' (Madrid City Council, 2007, p. 3). This situation has led to the neglect by local authorities and the consequent deterioration of buildings. These aspects framed a discourse that focuses on the decline of traditional methods of food and service provision.

It was in this context that there were the first talks of crisis and the need to renovate retail markets – even though markets were the leading channel for the distribution of fresh food. The response of public authorities was the elaboration of an action plan to 'reactivate' markets. This initiative involved the selling of stalls to their own shopkeepers for direct administration purposes. This strategy concealed a purely economistic and short-term approach that initiated the redefinition of markets as commercial and profitability assets. Madrid City Council has presented the changes in consumption patterns in the city and the decline of public markets as a natural evolution in retail processes. And although in their publications they display a rhetoric of markets as a 'public service embedded in the customs of the Madrid citizens' (Madrid City Council, 2010, p. 1) a careful and critical analysis of their policies reveals that they have sought to distance themselves from the direct management of markets leading to the weakening of the traditional and affordable market model (Mateos, 2017)

The local authority's plan to sell market stalls directly to the traders was only applied on one occasion, and so this situation begs the development of different hypotheses. It could have been too soon for investors to take part in renovation processes and see markets as new and profitable sources of investment; or that there was still a need to reach what Neil Smith describes as a 'rent-potential gap' (Smith, 1996) to ensure returns from reinvestment for renovation purposes. As for the key factor that would enable the achievement of this initiative, there would be a need of an 'interventionist' policy and the definition of proper regulatory frameworks for the reinvestment of capital in the form of subsidies for urban renovation projects or the entry of new investors into municipal markets. In other words, there was a need to define public–private partnerships capable of absorbing intervention risks within the context of the so-called 'entrepreneurial urbanism' (Harvey, 1989).

These conditions or 'latent potentialities' were gradually defined by urban regulation (Madrid City Council, 1997) and the agreements signed among different public entities (the central state, the regional authorities and the City Council), thus giving rise to a series of urban renewal policies designed to intervene in the historical centre of Madrid. During the 1980s, most environmental, economic and social improvement strategies focused on the local sphere, the neighbourhood and its dwellers; such an approach, however, progressively evolved into the adoption of a 'systematic and deliberate policy that favoured the global dimension over the local domain' (Justo, 2011, p. 74). This action meant the commodification of the urban space and its residential, cultural, recreational and commercial elements and the renovation/expulsion of local dwellers. Perhaps the clearest example of this issue is the Malasaña-Chueca-Hortaleza-Gran Vía area (including the 'new' Mercado de San Antón and Mercado de Barceló and the now-threatened Mercado de los Mostenses).

This environment led to the emergence of a new municipal policy on markets, which was based on the redefinition of strategic and regulatory frameworks. All recently-launched initiatives were presented by the authorities as an opportunity for markets to adapt themselves to the current retail trends in

Madrid, which are dominated by large and corporate players and shopping centres. In order to promote competitiveness, three strategies were designed: the renovation and modernisation of markets, the professionalisation of managerial activities and the improvement and promotion of a common image through urban marketing campaigns. Since supermarkets were regarded as 'commercial formats perfectly adapted to the new pace of life', markets in Madrid had to introduce themselves as 'new, original and different commercial formats that maintain the spirit and characteristics that make markets into unique commercial elements because of their neighbourly and familiar atmosphere' (Madrid City Council, 2011, p. 1).

Finally, despite the fact markets and supermarkets are described as differentiated formats, municipal plans aim at incorporating the latter into the former for two purposes: they want supermarkets to attract customers and they want to make them bear some of the transformation costs. Parallel to this, the new municipal markets policy aims to provide market managers with greater management autonomy and redefine land use through a considerable decrease in the minimum percentage that needs to be designated for the selling of fresh food (from 65 per cent to 35 per cent). It did not take long for the first measure – which conceals within it a privatisation measure – to take effect. In 2013, new private commercial entities (owned by construction or real estate companies) were awarded contracts to manage some markets, thus assuming a role once played by the Association of Traders. In their search for profit, these new management companies applied measures to sharply increase the rents for market stalls (Maiello, 2014). On the other hand, the new flexible rules on the types of trade allowed in markets has enabled the establishment of supermarkets and other types of commercial activities (such as bars, restaurants or offices).

Inequality and uneven developments within the context of markets in Madrid

Fragmentation and inequalities in the types of markets

As we have been stressing in this book, the contested nature in cities is also reflected in markets. On the one hand, there is a neoliberal-oriented model that seeks to make private gains from markets and, on the other hand, there are some community action and self-management initiatives that provide new approaches to food sovereignty, local economies, etc. In the case of Madrid, there is also a tension between these diverse trends affecting the already heterogeneous network of markets and thus it is impossible to talk about the emergence of a single market 'model'.

We have conducted an analysis with data across markets in Madrid of the number and type of stalls, renovation status of markets and the type of offer they have and we have been able to identify four types of trends of markets, which are described below:

- Traditional markets, characterised by the provision of essential, fresh food products.
- Gourmet markets, where 10 per cent of stores are engaged in the sale of *delicatessen* products, which are targeted at the restaurant business.
- Organic markets, where 10 per cent of stores are engaged in the sale of healthy, fair-trade organic products.
- Service-oriented markets, where there are no traditional stalls to acquire fresh produce and where more than 75 per cent of stores are engaged in the provision of non-food services (hairdressers, clothing and shoe shops, etc.).

Understanding the heterogeneity of models: relationships between geographical location and typologies

When we translate this information into a map, this shows how the different types of markets are distributed according to well-defined geographic patterns that reveal a conflict between the two socio-spatial realities of Madrid: the centre vs the periphery. This contrast is revealed in the fact that the central districts of Madrid (Centro, Retiro, Salamanca and Chamberí) have a per capita income one and a half times higher than the districts located in the southern periphery area (Latina, Carabanchel, Usera, Villaverde and Puente de Vallecas) (Madrid City Council, 2009) – not to mention that these districts are separated by the M30 ring road. It is not therefore surprising that gourmet and eco-markets are exclusively located in the central area of the city (Figures 7.1a and b). On the other hand, service-oriented markets are scattered throughout peripheral areas and traditional markets are distributed all over the city (Figures 7.1c and d). In any case, it is difficult to find unique models, since in both central and peripheral areas markets are showing hybridisation trends.

Markets and the double gentrification frontier

The analysis of Madrid shows that the transformation of markets involves the emergence of a double gentrification border. The first of these boundaries is located within markets, where shopping activities are altered to the point of turning shopping into an exclusive experience that can only be accessed by those who can afford the gourmet products found in specialised markets. The segregation generated by these commercial gentrification processes, which are based on the consumption of exclusive goods, is not only derived from the unequal access to the goods and practices offered by markets. There is also a type of segregation associated with culture, which operates in the same way as the exclusionary practice described above, and is an essential part of it. In this sense, gourmet consumption practices are reflective processes based on the controlled development of taste and judgment (Crewe and Beaverstock, 1998; Featherstone, 1991; Lash and Urry, 1994) that cannot be equally accessed by people. Therefore, the targeting of an increasing 'food elite' (Holloway and

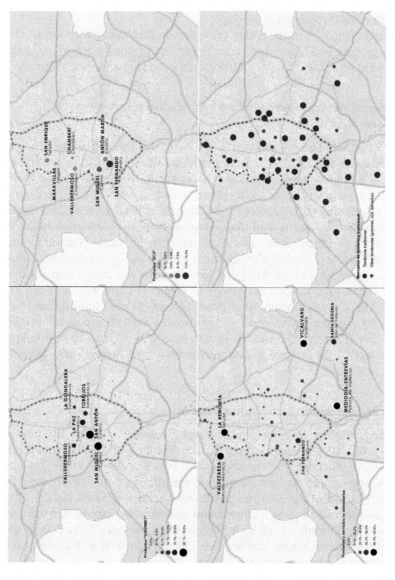

Figure 7.1 Different typologies of markets. Top left (a): Gourmet; top right (b): organic; bottom left (c): Service; bottom right (d): traditional

Source: Alejandro Rodríguez.

Kneafsey, 2000; Hurst, 1998) has become one of the pillars of segregation and one of the reasons through which markets start to lose their public service functions.

The second border is not as abstract as the previous one; this is a physical frontier. Since the central districts of Madrid are mainly focused on commercial specialisation, peripheral areas are not included in the transformation processes that take place in the centre of the city. In this sense, investors know where to allocate their capital; this can be seen in the different levels of renovation experienced by markets in Madrid. The only renewal initiatives implemented in peripheral markets are associated with minor structural adjustments; however, the central area of the city has undergone comprehensive renovation processes and seen the emergence of new markets (in this case projects were funded 50 per cent by the local administration and 50 per cent by the private sector).

In line with this research framework of frontiers, in the next sections we analyse three case studies on markets which, despite being located in the same area (Centro district), show different realities and situations within the heterogeneous context of markets in Madrid

Mercado de San Antón: the triumph and exhaustion of a role model

Mercado de San Antón is located in the Chueca neighbourhood of central Madrid, along Gran Vía. This was a low and middle income neighborhood during the first half of the twentieth century; however, the 1970s witnessed the beginning of a progressive deterioration process that led to the loss of population, abandonment of dwellings and social issues. During the early 1980s, this area was the epicentre of 'la Movida', a counter-cultural movement that took place in Madrid since the late 1970s, then emerging as a gay neighbourhood in the 1990s. Chueca is a clear example of gentrification and today is one of the most expensive neighbourhoods in Madrid, with a focus on recreation and leisure activities (García Pérez, 2014).

This area is home to Mercado de San Antón, whose first building was built in 1945. With no architectural aspirations, this space operated for decades as a neighbourhood market for the dwellers of the popular Chueca neighbourhood until the 1980s and 1990s, a period of time marked by great social change. At the end of the twentieth century, the old Mercado de San Antón had about 50 stores that offered traditional everyday food products.

In 2002, with the goal of modernising the building and its facilities, the Association of Traders decided to take part in the Modernisation Plan promoted by the City Council. Though 35 per cent of the renewal project was funded by the local administration, traders were not able to afford such expenditure. Throughout the comprehensive renovation process (the most expensive intervention within the municipal programme) more than half of traders relinquished their concession rights in exchange for compensation (funded by El Corte Inglés, a national chain of department stores), while others rented their shops to food and restaurant franchises.

The new Mercado de San Antón opened in May, 2011, and became the leading gourmet space in the city, following in the footsteps of the renowned Mercado de San Miguel (an old private market built in the nineteenth century that was transformed into the pinnacle of delicatessen cuisine). The original market had 48 stores; today this space has been divided into 12 'traditional' shops, 10 bar-restaurants, a terrace-restaurant area, an exhibition space, a supermarket and an underground car park. Mercado de San Antón has clearly followed the gentrification trends of its home neighbourhood, where the purchasing experience, bars and restaurants and gourmet products are key elements.

However, from our perspective, the model used by Mercado de San Antón – a prime example of neoliberalism – has shown signs both of victory and of exhaustion. Victory because it has set a local and international trend and attracted new customers and exhaustion because despite the aspiration many other markets have not adopted this gourmet model. In any case, the existence of this type of market has put in question the viability of the more traditional types of markets and their nature as public services.

Figure 7.2 Gourmet mushrooms on offer at San Antón Market
Source: Rodríguez Sebastián, 2012.

Mercado de los Mostenses: A non-articulated resistance

Mercado de los Mostenses is located in one of the main streets of the capital: Gran Vía. This is a rather undefined area without a clear function either residential or service-related, historically characterised by a certain level of marginalisation, this space has welcomed the arrival of a migrant population. Since the 1990s, the area has been the target of speculative pressure as the result of the comprehensive renewals processes experienced by adjacent neighbourhoods – particularly in the case of Malasaña – and the private commercial project 'Triball S.L.'.

The old Mercado de los Mostenses (a beautiful building made of glass and iron) was inaugurated in 1875 and torn down in 1925 as the result of the construction of Gran Vía. The current building, which is made of brick, was completed in 1946 and operated as a local market for decades. During the 1990s, the crisis of this retail format in the city prompted Mercado de los Mostenses to focus on the new migrant communities (mostly from Latin America and Asia) that inhabited this area of Madrid. With its 100 stores specialised on the provision of traditional food items, the market became an 'emblem of social and commercial intercultural exchange in the city of Madrid' (Robles, 2012, p. 111).

However, despite this commercial success, Los Mostenses was not included in the Modernisation Plan promoted by the City Council. According to this authority, the market had to be demolished. In 2010, an architectural proposal for the construction of a multipurpose building was issued; however, such a project was never carried out. Then, in 2013, the private real estate developers Triball Group submitted a proposal for the renewal of the market that included the construction of two additional stories and a green roof. The firm would invest €6m in exchange for the administration of the new spaces.

As of today, Mercado de los Mostenses has not undergone significant intervention processes; this may be due either to the current economic situation, which has affected the financing of major investment projects, or to what we have called a 'non-articulated resistance' by the traders themselves which has emerged as a result of the mismanagement of the potential renovation project and passivity on the part of relevant actors. However, the market is still open and is deeply rooted within the neighbourhood; it operates at almost full capacity and offers a wide array of products targeted at users with low purchasing power.

Mercado de San Fernando: An experimental alternative

Surrounded by the Embajadores and Lavapiés neighbourhoods, the Mercado de San Fernando is located in the historical centre of the city. Initially, this popular district received migrants from rural zones in Spain and then during the 1980s the area experienced the arrival of newcomers from countries such as China, Morocco, Ecuador, Bangladesh and Pakistan. The multicultural nature of this area and the affordability of dwellings have attracted a diverse population of young students, activists and professionals, thus turning the Lavapiés neighbourhood into one of the most affordable housing options in central Madrid.

Mercado de San Fernando opened its doors in 1945 in an area once used by an open-air market. This monumental building, common in the Franco dictatorship era, contrasted with the very dense housing type of Lavapiés and operated as a neighbourhood market for decades. It was not until the 1990s that this market failed to properly respond to the demographic changes experienced by the neighbourhood. In this sense, while new migrants and young professionals prefer to buy their goods from small grocery stores or supermarkets, local and more elderly neighbours remained as the only users of this market.

In 1998, this space experienced an external and internal renewal process; however, such a measure failed to improve the economic activity of the market. Then, in 2005, the Association of Traders promoted an ambitious municipal renovation project that was never carried out as the result of funding issues on the part of local traders. 2007 saw the publicly-funded construction of a primary healthcare centre on the second floor of the market. This led to the idea of a series of possibilities for the construction of a supermarket; however, none of these ideas has met with success. By 2009, 40 per cent of the 50 shops that make up the market were closed.

At the end of that year, the market decided to take an alternative approach to that proposed by the City Council and large private operators. A group of traders opened talks with alternative collectives in the neighbourhood, experts and activists in order to embark on a bottom-up renovation of the market. In 2012, the rental cost of shops was published in an effort to recruit new traders and the 25 previously empty stalls were rented out at affordable prices, thus allowing the emergence of young and local-oriented initiatives. By the beginning of 2013, the market operated at full capacity and it was possible to find new activities such as organic food, crafts, professional services, a bookshop and home-cooked food. These businesses deal with local producers and aim to attract local middle class customers.

However, the incorporation of new restaurant and specialised food businesses suggests the need to seek the delicate balance between these activities and the traditional provision of fresh, everyday food products. This situation has become even more unstable in places such as Lavapiés, which has been closely observed by gentrification agents for years. At the moment, Mercado de San Fernando can be regarded as an alternative retail proposition (to big supermarkets and shopping malls) that regenerates the socio-commercial structure of the neighbourhood. However, if it ends up providing only bar and restaurant services, it could be assumed that the alternative model proposed by this market has failed to deliver the expected results. Therefore, the entry of businesses focusing on services other than the provision of fresh, essential food products can involuntarily encourage the proliferation of restaurant-oriented activities. Both types of these activities (service related and bars and restaurants) are complementary to each other as they operate under the same leisure-based consumption model, thus sharing the same opening hours – including late working hours and weekends. If this model continues to consolidate, the trade in essential goods may be seriously affected.

Conclusions

This research provides an overview of the transformations experienced by markets in Madrid. Within a context marked by the structural socioeconomic changes triggered by globalisation and the general decline in neighbourhood trade activities, we have paid special attention to the role played by municipal public policies in the renewal of markets. Public policies have given priority to the power and interests of some actors and ignored the presence of other relevant players. Modernisation has not therefore always ensured the survival of small-scale businesses.

As a result, these transformations have given rise to heterogeneous patterns associated with the specialisation and geographic location of markets (centre vs periphery). These aspects are explained by the close relationship that exists between markets and their respective local socioeconomic environments. Therefore, the current processes have led to the emergence of a selective and uneven transformation phenomenon framed within trends of commercial gentrification patterns. This is where the double gentrification border appears: the transformation of the shopping activity into a selective and unique experience and the real estate opportunities involved in the renewal of these central spaces.

These three case studies have enabled us to describe the main challenges arising from the transformation trends of markets, which are marked by tensions between neoliberal and grassroots/alternative visions for the city. In this context, at least three situations have been identified: the stagnation and passive resistance of the traditional model, which allow it to survive in areas where investment opportunities for urban operators may not yield the expected economic results; the progressive gentrification of markets through the gourmetisation of an already-gentrified residential environment; and the emergence of models that propose new forms of management and an alternative offer to traders and users. In conclusion, while markets are facing a renovation-or-die situation, their evolution may affect the coexistence of different lifestyles and social classes which, through consumption, ensure the right to the city.

Note

1 Original in Spanish translated by Juan Pablo Henríquez Prieto

References

Authier, J.-Y. (1989). Commerces et commerçants d'un espace en mutation. Le quartier Saint-Georges à Lyon. *Revue de géographie de Lyon*, 64(2), 63–69.

Bahamonde Magro, Á. and Fernández García, A. (2008). La economía: actividades económicas y mercado urbano. In: Comunidad de Madrid (Ed.), *Madrid, de la Prehistoria a la Comunidad Autónoma*. Madrid: Consejería de Educación, Secretaría General Técnica, pp. 475–497.

Barcelona City Council (2015). Mercados: La experiencia de Barcelona. Available from: http://ajuntament.barcelona.cat/mercats/sites/default/files/pub_mercats_CAT_def.pdf

Berroir, S., Clerval, A., Delage, M., Fleury, A., Fol, S., Giroud, M., Raad, L. and Weber, S. (2015). Commerce de détail et changement social urbain: immigration, gentrification, déclin. *EchoGéo*, 33. Available from: https://echogeo.revues.org/14353

Brenner, N., Peck, J. and Theodore, N. (2010). Variegated neoliberalization: geographies, modalities, pathways. *Global Networks*, 10(2), 182–222.

Bridge, G. and Dowling, R. (2001). Microgeographies of retailing and gentrification. *Australian Geographer*, 32(1), 93–107.

Chabrol, M. (2011). *De nouvelles formes de gentrification? Dynamiques résidentielles et commerciales dans le quartier de Château-Rouge (Paris)*. Thesis. Paris: Université de Poitiers.

Checa, A. and Lora-Tamayo, G. (1993). El comercio minorista de Madrid. *Espacio, Tiempo y Forma, Geografía, Serie VI*, n.6, 79–138.

Crewe, L. and Beaverstock, J. (1998). Fashioning the city: Cultures of consumption in contemporary urban spaces. *Geoforum*, 29(3), 287–308.

Featherstone, M. (1991). *Consumer Culture and Postmodernism*. London: Sage.

Fleury, A. (2003). De la rue-faubourg à la rue 'branchée': Oberkampf ou l'émergence d'une centralité des loisirs à Paris. *Espace Géographique*, 32, 239–252.

García Pérez, E. (2014). Gentrificación en Madrid: de la burbuja a la crisis. *Revista de Geografía Norte Grande*, 58, September, 71–91.

González, S. and Waley, P. (2013). Traditional retail markets: The new gentrification frontier? *Antipode*, 45(4), 965–983.

González, S. (2011). Bilbao and Barcelona 'in motion'. How urban regeneration 'models' travel and mutate in the global flows of policy tourism. *Urban Studies*, 48(1), 1397–1418.

Guàrdia, M. and Oyón, J.L. (2007). Los mercados públicos en la ciudad contemporánea. El caso de Barcelona. *Biblio 3W, Revista Bibliográfica de Geografía y Ciencias Sociales*, XII(744). Available from: http://www.ub.es/geocrit/b3w-744.htm

Guàrdia, M. and Oyón, J.L. (Eds.) (2010). *Hacer ciudad a través de los mercados. Europa, siglos XIX y XX*. Barcelona: Museu d'Historia de Barcelona / Institut de Cultura, Ajuntament de Barcelona.

HackworthJ. and RekersJ. (2005). Ethnic packaging and gentrification: The case of four neighborhoods. *Toronto Urban Affairs Review*, 41, 211–236.

Harvey, D. (1989). From managerialism to entrepreneurialism: The transformation in urban governance in late capitalism. *Geografiska Annaler. Series B, Human Geography*, 71(1), 3–17.

Harvey, D. (2008). *París, capital de la modernidad*. Madrid: Akal.

Hernández Cordero, A. (2014). Gentrificación comercial y mercados públicos: El mercado de Santa Caterina, Barcelona. Working Paper Series, Contested Cities,

Series I., 14017, May. Available from: http://contested-cities.net/working-papers/2014/gentrificacion-comercial-y-mercados-publicos-el-mercado-de-santa-caterina-barcelona/#comments

Holloway, L. and Kneafsey, M. (2000). Reading the space of the farmers' market: A preliminary investigation from the UK. *Sociologia Ruralis*, 40(3), 285–299.

Hurst, C. (1998). Food, glorious food. *The Independent*, 11 November 1998.

Justo, A. (2011). Transformaciones en el barrio de Malasaña. Hacia la gentrificación. *Viento Sur*, 116, 73–79.

Lash, S. and Urry, J. (1994). *Economies of Signs and Space: After Organised Capitalism.* London: Sage.

Lees, L., Slater, T. and Wyly, E. (2008). *Gentrification.* Abingdon, UK: Routledge.

Lehman-Frisch, S. (2002). Like a village: les habitants et leur rue commerçante dans Noe Valley, un quartier gentrifié de San Francisco. *Espaces et Sociétés*, 108–109, 49–68.

Lemarchand, N. (2011). Nouvelles approches, nouveaux sujets en géographie du commerce. *Géographie et cultures*, 77, 9–24.

López de Lucio, R. (Dir.) (1998). Espacio público e implantación comercial en la ciudad de Madrid. Calles comerciales versus grandes superficies. *Ci[ur], Cuadernos de Investigación Urbanística*, 23. Retrieved: http://polired.upm.es/index.php/ciur/article/view/237/233

López de Lucio, R. (2007). Comercio y periferia: el caso de la región de Madrid. *Ciudades*, 10, 185–202.

Madrid City Council (1997). *Plan General de Ordenación Urbana de Madrid.* Madrid: City Council.

Madrid City Council (2007). *Plan de Innovación y Transformación de los Mercados Municipales de Madrid, 2003-2007. Cuatro años impulsando el comercio.* Madrid: City Council. Available from: http://www.madrid.es/UnidadesDescentralizadas/Consumo/ficheros/libromercados.pdf

Madrid City Council (2009). *Contabilidad Municipal de la Ciudad de Madrid.* Dirección General de Estadística, Área de Gobierno de Hacienda y Administración Pública.

Madrid City Council (2010). Ordenanza de Mercados Municipales ANM 2010/62. *Boletín Oficial del Ayuntamiento de Madrid*, 6.340, 30 December, pp. 7–28.

Madrid City Council (2011). *Plan de Innovación y Transformación de los Mercados Municipales de Madrid, 2003–2011. Ocho años impulsando el comercio.* Madrid: City Council.

Maiello, V. (2014). El mercado de los mercados: Análisis de los procesos de transformación de los mercados municipales de abastos de Madrid. Working Paper Series, Contested Cities. Series I. WPCC-14016. Available from: http://contested-cities.net/working-papers/2014/el-mercado-de-los-mercados-analisis-de-los-procesos-de-transformacion-de-los-mercados-municipales-de-abastos-de-madrid/

Martin, N. (2014). Food fight! Immigrant street vendors, gourmet food trucks and the differential valuation of creative producers in Chicago. *International Journal of Urban and Regional Research*, 38(1), 1867–861883.

Mateos, E. (2017). Transformación del comercio de proximidad: legitimidad y disputas. *Revista Ciudades*, 114, 10–16

Mermet, C. (2011). Redéfinir la consommation pour repenser les espaces de consommation. *Géographie et Cultures*, 77, 25–44.

Michaud, Y. (2015). *El nuevo lujo.* Barcelona: Taurus.

114   *Eva García Pérez et al.*

Montoliú Camps, P. (1988). *Once siglos de mercado madrileño: de la plaza de la Paja a Mercamadrid*. Madrid: Sílex Ediciones.

Omhovère, M. (2014). Vivre à Gambetta, l'ancrage local. Lorsque la proximité devient ressource. In: Fol, S., Miot, Y. and Vignal, C. (Eds.), *Mobilités résidentielles, territoires et politiques publiques*. Lille: Presses du Septentrion, pp. 229–248.

RankinK.N. and McLeanH. (2015). Governing the commercial streets of the city: New terrains of disinvestment and gentrification in Toronto's inner suburbs. *Antipode*, 47(1), 216–239.

Robles, J.I. (2012). El Mercado de los Mostenses (Madrid). *Distribución y Consumo*, 122, March–April, 111. Available from: http://www.mercasa.es/files/multimedios/1336048490_pag_110-114_Mostenses.pdf

Rodríguez Sebastián, A. (2014). La transformación de los mercados municipales de Madrid. Análisis legislativo, comercial y económico de los mercados de abastos madrileños. *Territorios en Formación*, 7, 7–8.

Roth, N. and Grant, J. (2015). The story of a commercial street: Growth, decline, and gentrification on Gottingen Street, Halifax. *Urban History Review*, 43(2), 38–53.

Schwentesius, R. and Gómez, M. (2006). Supermercados y pequeños productores hortofrutícolas en México. *Comercio Exterior*, 56(3), 205–218.

Semi, G. (2005). 'Chez Saïd' à Turin, un exotisme de proximité. *Ethnologie française*, 35(1), 27–36.

Shkuda, A. (2013). The art market, arts funding, and sweat equity: The origins of gentrified retail. *Journal of Urban History*, 39(4), 601–619.

Smith, N. (1987). Gentrification and the 'rent gap'. *Annals of the Association of American Geographers*, 77(3), 462–465.

Smith, N. (1996). *The New Urban Frontier. Revanchist City and Gentrification*. New York: Routledge.

Smithers, J., Lamarche, J. and Joseph, A.E. (2008). Unpacking the terms of engagement with local food at the farmers' market: Insights from Ontario. *Journal of Rural Studies*, 24(3), 337–350.

Sullivan, D. and Shaw, S. (2011). Retail gentrification and race: The case of Alberta Street in Portland, Oregon. *Urban Affairs Review*, 47(3) 413–432.

Van Criekingen, M. and Fleury, A. (2006). La ville branchée: gentrification et dynamiques commerciales à Bruxelles et à Paris. *Belgeo. Revue belge de géographie*, (1–2), 113–134.

Varanda, M. (2005). La réorganisation du petit commerce en centre-ville. *Revue française de sociologie*, 46(2), 325–350.

Viana Cerqueira, E.D. (2014). A evolução das formas de gentrificação: estratégias comerciais locais e o contexto parisiense. *Cadernos Metrópole*, 16(32), 417–436.

Watson, S. (2009). The magic of the marketplace: Sociality in a neglected public space. *Urban Studies*, 46(8), 1577–1591.

Zukin, S., Trujillo, V., Frase, P., Jackson, D., Decuber, T. and Walker, A. (2009). New retail capital and neighbourhood change: Boutiques and gentrification in New York City. *City & Community*, 8(1), 47–64.

8 Mercado Bonpland and solidarity production networks in Buenos Aires, Argentina

Victoria Habermehl, Nela Lena Gallardo Araya and María Ximena Arqueros

Introduction

This chapter uses Mercado Bonpland, a small market organised by many self-managed organisations in Buenos Aires, as a focus for how alternative economic practices such as economic solidarity, alternative consumption, production and exchange take place. Mercado Bonpland demonstrates the many solidarity practices, both in terms of economic and collaborative organisation, that can take place in such spaces. The value of this chapter is that it highlights these solidarity practices, which also exist in other markets, such as those discussed in this book. In order to identify these solidarity practices, the introduction develops the history of solidarity and social economy organising in Buenos Aires. In the second section, Mercado Bonpland operates as an important bridge linking production and consumption places, practices and collectives. In the last section we explore the way that urban to rural production and consumption are connected to national and international networks. In this way the market demonstrates the complexity of networks of solidarity economies beyond the market space itself. Our methodological approach encompasses different qualitative research into the study of markets and production conditions. This research is based on interviews and ethnographic research in Mercado Bonpland as well as with producers outside of the market.

In this chapter we argue that the creation of alternative economy spaces in Buenos Aires is established through rural–urban networks of producers and consumers. These networks were strengthened and generated as a way to respond to the multiple crises of 2001 in Argentina (economic, political and social). It is worth saying that the crisis was not confined to one event but was long term, and can be traced to the state withdrawal of the 1990s, which also compelled people to take politics into their own hands.

By the late 1990s, the feasibility of neoliberal development was criticised in academic and policy circles, which led to the creation of alternative social policies. These included the Social and Solidarity Economy (Hintze, 2007a, p. 4) that occupied an important role and would later be taken on as state policy by subsequent progressive left governments. The Social and Solidarity Economy proposal centres around the idea developed by Coraggio and Sabaté that 'another economy is possible' based on the principle of social inclusion. This built on an analysis of the difficulty of reintegrating most of the population through Keynesian economic policies (Coraggio, 1999 and Federico-Sabaté, 2003 in Hintze, 2007b). After the crisis in Argentina, most people were excluded from the economy, which meant that individual citizens had to find their own solutions whilst some social policies were also implemented to mitigate the consequences of neoliberal model.

At the same time, a new movement of neighbourhood assemblies emerged as experiences of self-management that not only resisted these neoliberal economic models but created 'alternatives' through direct action (Fernández, 2011). In time these were replicated in other areas with similar characteristics as well as their own particularities, such as urban gardens, cultural centres, public libraries and worker managed factories. Thus, across different scales there were hundreds of experiences of self-management; some were supported by public programs, whilst others were autonomous and tried to solve social exclusion through integrating more people into production (Coraggio and Sabaté in Hintze, 2007b). These were organised through different groups such as community associations, organised production of groups of unemployed workers, mutual aid societies, self-managed public services, NGOs and private foundations utilising volunteer work, craft workshops, microenterprises, cooperatives, mutuals, bankrupt companies managed by their workers, solidarity barter markets, initiatives of social reproduction and self-employment of domestic units in the city, solidarity credit, and so on. They were also supported by researchers at public universities (ibid.). These new forms of economic organisation brought about tensions between two different approaches: firstly, that of autogestion, direct democracy, horizontal organisation and, secondly, vertical structures of organisation such as bureaucratic processes or organisation through the state (Fernández, 2011 p. 130). The results of these projects were often hybrid combinations of both approaches; for example, cooperatives that function in assemblies in opposition to the state, yet rely on them to facilitate certain conditions of production. We situate our case study of the Mercado de Economia Solidaria Bonpland (Bonpland Solidarity Economy Market) as one such experience that connects many examples of these diverse forms of organisation. Mercado Bonpland is therefore one case in the multiple and diverse examples of self-managed practices in Buenos Aires but more widely can also be seen as an illustration of collective practices that are already happening in other markets across the world.

Mercado Bonpland opened in 1914 as part of a network of 36 municipal markets that were built in Buenos Aires between 1856 and the beginning of the twentieth century as the result of a public policy to provide food for a growing urban population (Medina and Álvarez, 2009). Mercado Bonpland is located in the northern neighbourhood of Palermo, in a district known as Palermo Hollywood, an area that has been transformed radically by investments leading to the area being full of bars, restaurants and TV studios and leading to a displacement of previous residents from the neighbourhood

(Herzer et al., 2015). During this urban transformation the municipal market was slowly abandoned, particularly during the late 1990s, until 2007 when the Palermo Viejo neighbourhood assembly, which emerged during the 2001 crisis mentioned above, organised and moved inside the market building (Mauro and Rossi, 2013). Mercado Bonpland as we know it today consists of 17 stalls each run by a different cooperative association, producer or small network. We focus particularly on the food described as 'natural', 'organic', 'healthy' and 'direct from producers' that are characterised by various actors who integrate rural–urban networks.

Mercado Bonpland: The reconstruction of the networks

In the district of the City of Buenos Aires (Ciudad Autónoma de Buenos Aires, CABA) we find, as in other mega-cities across the world, the abandonment of traditional retail markets in favour of shopping centres of mega consumption (e.g. Abasto or Spinetto shopping centres). As with other cities discussed in this book, gentrification of some of these traditional spaces is mainly driven by tourism and gourmetisation trends as demonstrated in the proliferation of 'organic' markets in the neighbourhoods of Palermo and San Telmo. Mercado de Economia Solidaria Bonpland (Palermo, since 2007) and El Galpón Market (in Chacarita neighbourhood, since 2005) also sell 'organic' products but with a commitment to a solidarity economy and agroecology. They are characteristic of solidarity economy markets as they engage in direct social relationships between producers and consumers on the basis of removal and/or elimination of intermediaries; fairer prices; healthy products; lack of worker exploitation; promotion of gender equality; environmental protection and so on. However, these markets are not totally autonomous, as although there are many different experiences, their organisation takes place through co-operatives and worker-led companies, some of which have state support to continue (Caracciolo, 2014).

The Government of the City of Buenos Aires also influences markets and fairs through its public policies. On the one hand, it subsidises Itinerant Neighbourhood Provision Markets, which are traditional municipal markets that are organised once a week in each neighbourhood of Buenos Aires. These markets offer fresh affordable produce for the middle and working classes through government-controlled prices. On the other hand, as part of a policy of reclaiming public space to build a 'green city' (the slogan of the government for Buenos Aires from 2007 to the present), gourmet markets such as Buenos Aires Market are promoted, especially to middle and upper class consumers, where fairs operate as elite weekend events. Simultaneously tourist markets linked to gastronomy, in squares, public parks and streets, promote the illusion of traditional, healthy and locally produced foods. In both of these cases Social and Solidarity Economy policies are not promoted. However, at the same time, the national government has also implemented policies to support the Social and Solidarity Economy. Consequently, these

policies generated different tensions; for example, some markets receive grants
and permission to use public spaces whereas others do not.

Overall, therefore, there is a multiplicity of experiences in terms of markets
in Buenos Aires. Yet, though the appearance of some solidarity markets may
seem similar to gourmet events, economic solidarity markets in Buenos Aires
demonstrate how markets can function as spaces through which to build
alternative networks. Resistance to 'business-as-usual' global economic processes can be seen in the resourceful, community-oriented, co-operative and
independently organised markets such as Mercado Bonpland. This case
highlights the potential of establishing alternative economic and social
production systems within all markets. In the following two subsections we
elaborate further on the roots of the Mercado Bonpland; namely, the rich
history of agroecology networks of production and consumption in Argentina
and the autogestive urban assemblies that can all be conceptualised within the
Social and Solidarity Economy.

The agroecological movement

Mercado Bonpland's producers organise around agroecological production
conditions as well as solidarity economic practices. It is important to distinguish
between production networks organised around 'agroecological principles'
from 'organic production'. Each have different principles: the focus of agroecological production is for producers to create more just production conditions, without intermediaries, towards achieving food sovereignty rather than
focusing only on creating a high value product. Moreover, agroecological
production is connected with markets and fairs that support alternative
exchange networks, often organised by social movements that emerged in
Argentina after the 2001 crisis. These practices are used to develop the Social
and Solidarity Economy. There has long been academic debate over the use
of the terms Social and Solidarity Economy in Europe and Latin America
from social movements (Coraggio, 2002; Defourny, 2003; Laville et al., 2007;
Singer, 2010). Yet, the focus of the Social and Solidarity Economy is to
include those excluded from the broader economy by creating an active
relationship with producers.

Additionally, Mercado Bonpland must be situated within the long history
in Argentina of alternative economic networks demonstrated through ferias,
markets and particularly in the form of *ferias francas* (Carballo, 2000; Carballo et al., 2004; Goldberg, 1999; Sevilla Guzmán and Martínez Alier, 2006).
These fairs have been part of the process of reclaiming production rights for
many years, and are facilitated through non-monetary exchange. This
exchange has provided a way of moving beyond producing cash crops for the
international market and instead supports small-scale independent production
(García Guerreiro, 2014). Furthermore, these ferias are important in resistance histories, facilitating barter, production for survival, and against land
grabs and big producers.

At the present time many of these ferias ascribe to the principles of the Social and Solidarity Economy and connect with 'new rural social movements' emerging across the world in defence of traditional agroecological methods (Sevilla Guzmán and Martínez Alier, 2006). These movements are born out of local resistance to multinationals producing seeds, degradation of ecosystems and the threats to livelihoods because of agricultural modernisation. They are based on ancient knowledge of farming systems and innovations of low input agriculture. As Sevilla Guzmán and Martínez Alier (2006) discuss, there is a difference between agroecological farming in Latin America and organic farming in Europe; agroecological movements in Latin America often function as advocacy for large rural populations, peasants and landless labourers. Importantly agroecological movements develop and support direct sales (such as in Mercado Bonpland), which rely on complex networks of rural and urban actors coordinated at the local, national and global level built over decades. An example of these networks developed through agroecological principles and sold in Mercado Bonpland is the yerba mate called 'Titrayjú'. The name of this Argentine tea is a combination of the words 'tierra, trabajo y justicia' (land, work and justice) which represents the focus of their work.

The history of Titrayjú (which is currently sold in Bonpland) is an example of the necessity of producers organising networks of consumption, relationships and the history of political struggles. Titrayjú is a product of a cooperative created in 1975, in the city of Oberá, by MAM (Movimiento Agrario Misionero, Agrarian Missionary Movement) to generate more potential for autonomous farming. During the military dictatorship (1976–1983), political persecution meant that the cooperative stopped production, but activities began again in 1985, with the change to a democratic government (Vázquez, 2006). In the context of the globalisation and corporate concentration of agricultural production (Teubal and Rodriguez, 2002), the cooperative started participating in *ferias francas* (described earlier) in Argentina and began to gain the support of public institutions such as PSA (Programa Social Agropecuario, Agricultural Social Program) (Vázquez, 2006). In 2001 the cooperative started the brand Titrayjú (ibid.) creating an alternative circuit to the traditional forms of commercialisation, which they developed through direct sales to consumers (ibid.). Through contacts with social organisations they made initial direct sales in Buenos Aires; however, it was only through fairs like the *feria de Mataderos*[1] that networks of direct buyers were created and future members developed, including Mercado Bonpland. Thus, Titrayjú is one case that demonstrates the history of organising through resistive exchange markets and how the fairs that became common during the 2001 crisis led to the development of networks of alternative economies.

The assembly movement

Whilst Titrayjú demonstrates the history of agroecological networks and organising, Mercado Bonpland is also a result of the assembly organising from 2001 in Argentina. The assemblies were the form of resistance in the urban environment to the neoliberal model from both the middle and working classes. The exchange of goods on the street at *ferias* was a key survival strategy in the crisis, as well as a way for people to independently organise their daily lives. Ana, a market stallholder and member of the Palermo Viejo Assembly in Buenos Aires explains how she was involved in fairs in urban public spaces. This experience led to her initial involvement in Mercado Bonpland:

> [I]t was a time when there were a lot [of fairs] – every fortnight, with people in San Telmo fair, from San Telmo Assembly, we put them in Diagonal Sur, where the Bank of Boston is and Florida. We would assemble in public space ... That was a big movement you see. And I always came to assemblies that were made here, and participated.
>
> (Interview with Market stallholder Ana, 2014)

As Ana describes, the movement of ferias in urban public spaces also provided a way for people to come together, and these fairs in prominent public spaces were thus used as a meeting point as well as a means for building connections. The ferias therefore went beyond just exchanging goods, to contributing to the organising of the neighbourhood.

Mercado Bonpland and the networks of autogestive projects that it relies on are organised through different scales and varying actors, but both are connected through the exchange of goods, knowledge and experience, amongst other things. As such, Mercado Bonpland underlines the potential to organise autogestive projects between and through different spaces. Highlighting the state's governance and control of public spaces, Mercado Bonpland focused on daily survival and producing solidarity economic practices.

Mercado Bonpland: A bridge in the network of solidarity economies

Mercado Bonpland as it is today was formed in 2007 in the abandoned traditional retail market (1914) (Medina and Álvarez, 2009) on Bonpland Street, in the gentrified neighbourhood of Palermo (see Figure 8.1). This small market operates at a completely different scale from many of the other markets that are featured in this book. However, as well as this small scale, Bonpland demonstrates an attempt to organise around a different set of principles than those that are the premise of many of the other markets. The case of Mercado Bonpland allows a focus on the organisation of alternative economic practices such as economic solidarity, alternative consumption, production and exchange. The focus of Bonpland on the Social and Solidarity Economy questions the logic of capitalist production and consumption on the premises of fair trade (fair prices to producers), autogestive or self-managed production, healthy food and responsible consumption.

Mercado Bonpland acts as a 'bridge' between initiatives (its aim is to facilitate connections between networks), as well as a space in the city for

Figure 8.1 The exterior of Mercado Bonpland.
Source: Victoria Habermehl, 2014.

face-to-face meetings. As such, the market is an organisational point for economic solidarity initiatives, as well as a visual symbol of these alternative
initiatives in the city and a method of facilitating different forms of consumption. A participant of La Asamblearia, one of the stalls in the market,
describes the importance of the market in connecting producers: 'Not intended
to be the centre of activity, so that each actor develops in its own way, [the
market] is only intended as a knot in that vast network, a bridge between
initiatives which now appear as isolated' (La Asamblearia, 2013). Bonpland
connects networks of alternative projects. The focus of the organisers is not
only to maintain and promote the marketplace itself but to help to improve

all of the projects associated with the network. The market is only important because of these many other projects, not despite them.

As already mentioned, the initial stimulus for the constitution of the Mercado Bonpland came from the Palermo Viejo neighbourhood assembly which started to meet inside the building in 2001. In particular, as part of their activities, the neighbourhood assembly organised 'La Trama' –a political-cultural festival held in May 2002, which developed economic solidarity initiatives as a focus for the neighbourhood assembly (Mauro and Rossi, 2013). This festival acted as an impetus for developing the theme of economic solidarity; developing connections, engagements and networks with small producers, ferias and autogestive projects. In this context, the participants of assemblies wanted to develop solidarity with and to support small rural producers and cooperatives who had difficulties linked with sales of their products. After several years of organisation and location changes, the economic solidarity market in Bonpland was formed.

'La Trama' signified a break from organising by a broader assembly to a focus on economic solidarity projects. As we have already highlighted, the Social and Solidarity Economy in Argentina is a government project as well as something organised by social movements. Accordingly, in wanting to make change as well as be seen as 'legitimate actors in the neighbourhood' neighbours organised economic solidarity projects (ibid.). For La Asamblearia, an organisation in Mercado Bonpland:

> Solidarity Economy is the intent that is made from different stakeholders to articulate the economic emergency response that the popular sectors are giving to the crisis, making them come together in an integrated subsystem or economic sector.
>
> (La Asamblearia, n.d.)

Building on this rich history of social mobilisation, Bonpland and the cultural centre behind it are now constituted through many organisations (see Figure 8. 2). These organisations have connections either through cooperatives, historic organisations such as the assemblies or La Trama festival, or through working collectively. Within the market they produce and sell different products, such as: handmade clothes from Soncko Argentino; vegetables from CEDEPO (Centro Ecuménico de Educación Popular or Centre for Promotion of Religious Unity and Popular Education) and APF (Asociación de Productores Familiares, Family Producers Association); and products from reclaimed factories as La Alameda. At the same time there are cooperatives that organise and sell agroecological products, such as La Asamblearia and Colectivo Solidario.

Furthermore, there are some cultural groups as Movimiento Popular La Dignidad (The Popular Movement for Dignity). Whilst there are overlaps, as many stalls support multiple projects (and many projects work across these categories), this emphasises the predominant organisational aims. Through their organisation these networks therefore bridge a divide between the perceived separation of urban/rural or production/consumption. Mercado Bonpland therefore demonstrates connections that these autogestive networks build through organisation between apparently different projects.

Consequently, in organising around this solidarity economy, the neighbours and now market organisers are seeking to develop more 'reliable' economic approaches, through improving individual and collective capacities to create the economy. This improves life in the neighbourhood, as well as connecting to many other spaces where similar projects are trying to create alternative economies. For Bonpland, this means developing relationships with these other spaces such as small agricultural projects, self-managed factories, cooperatives

Figure 8.2 Map showing layout of market, 2015
Source: Photo: María Ximena Arqueros.

and artisans. These groups prioritise self-managed or dignified work, fair trade and responsible consumption.

Mercado Bonpland operates as a space that not only Palermo locals rely on, but where autogestive projects and 'alternatives' to ordinary economic organising can sell their products. However, the physical space of the market is also vitally important to what the market achieves. In the neighbourhood of Palermo, Mercado Bonpland is a visual sign and connection point to the other alternative projects in the city. In visiting Mercado Bonpland, consumers immediately have connections to the other spaces it supports – though information, people and products. This means that physically it demonstrates the power of these alternative economic ideas. This space is also a place where people can meet face-to-face with others. Bonpland is hence also useful as a meeting space, or a place to connect with others who are working on or supporting similar projects. In this way, Bonpland – particularly when organising classes, talks, theatre events and so on – operates similarly to cultural centres, whereby the space is crucial to develop and expand these networks of alternative projects.

One example is the cooperative Colectivo Solidario whose aims are to sell agroecological products, 'educate urban consumers' and encourage thinking about consumption as a political act with slogans such as 'Hacé justicia por compra propia. Anímate a un consumo diferente' ('Make justice through your shopping: Try a different consumption'). With these aims, they provide information about agroecological products and arrange debates with critical academics, experts and producers. Therefore, through these associations Mercado Bonpland acts as a resource of support for other projects; it is a space that people know and can rely on. La Asamblearia is another cooperative stall at Bonpland – an organisation and an example of an urban network which has supported Mercado Bonpland from the beginning in accordance with the principles of the assembly movement. This cooperative began after the crisis of 2001 and works as an intermediary connecting urban consumers with rural producers. These rural producers form part of peasant and indigenous movements as Movimiento Nacional Campesino e Indígena (Argentine Peasant and Indigenous Movement). We develop more of the urban–rural relationships in the following section.

Whilst Mercado Bonpland is organised as to create a solidarity economy, there are many ways in which, through being prepared to create and tackle everyday economic issues, Mercado Bonpland is embedded within these struggles. For example, it attempts to challenge relationships of capitalist consumption whilst at the same time it relies on relationships of consumption to function as a market. It highlights the potential for organising through networks of alternative economies and connecting to markets. We are not claiming that Mercado Bonpland is outside of these complex and contradictory relationships. In fact, organising from this everyday life context means that the market organisers, generally intermediaries between producers and consumers such as Colectivo Solidario and La Asamblearia, have encountered many challenges, emphasising the complexity of such arrangements rather than attempting to establish that Mercado Bonpland is outside of capitalism.

Mercado Bonpland: Connecting urban and rural spaces

The history of the neighbourhood planning (focused on autogestion from organisers living within and outside the neighbourhood), is crucial to understanding the development of the market. It is also important to understand the agroecological food and agriculture networks connected through Bonpland. This is characterised by two examples, CEDEPO and APF, who sell in Bonpland and demonstrate the crucial work of connecting producers and consumers through the sale of fresh vegetables.

CEDEPO (Centro Ecuménico de Educación Popular or Centre for the Promotion of Religious Unity and Popular Education) is a civil association and one of the main providers of the fresh vegetables in Mercado Bonpland. CEDEPO has a centre for education, research and ecological production called La Parcela, located in the rural–urban outskirts of Greater Buenos Aires known as Florencio Varela, a deprived area with high levels of poverty. In La Parcela, family farmers and experts produce food through a popular education and agroecological approach, and sell directly in Bonpland.

However, the CEDEPO project has a scope beyond food production. Elsa, a CEDEPO member and worker in Mercado Bonpland, explained that local people were initially encouraged to visit La Parcela to use the healthcare centre, as there were inadequate facilities in the area. When using the healthcare centre, people would see other things being produced which, in turn, built their interest in engaging with CEDEPO. The development of this organisation led to a significant improvement in people's basic daily lives. In talking to Elsa (and other CEDEPO members who work in Mercado Bonpland) she explained that previously local residents 'had no orchard, no hens, not even rabbits. All that appeared with CEDEPO' (Interview market stall holder Elsa, 2013).

Initially, La Parcela's educational role facilitated practical skills, such as how to produce agroecologically with the land using different techniques, such as permaculture, biodynamic and other traditional (peasant and indigenous) knowledges. The aim of this was to ensure that people with access to land also had the opportunity to learn skills that would help them produce, as producing their own food meant more self-reliance. The project was tailored to the needs of local residents, focusing on different forms of production and on improving local people's day-to-day lives.

These educational and community projects led to many new co-operatives and family agriculture projects, which have eventually become independent:

> Yes, it changed a lot, but also new co-operatives were born, like for example APF (Asociación de Productores Familiares – Family Producers Association). Neighbours started to get some information, they started to

build their own orchards, their own experiences, and so they started to create their own co-operatives.

(Interview with market stallholder Elsa, 2013)

This APF cooperative also produces vegetables which are sold in Mercado Bonpland, and although it began through taking part in CEDEPO, it now maintains independence. CEDEPO and its education and production focus has thus created a network of self-organised co-operatives that produce agricultural products. The independence of these groups ensures that people can produce and plan for themselves, and thus can improve their lives, rather than focusing on strengthening the organisation, CEDEPO, itself. Similar to Bonpland market, the focus of CEDEPO is that people who work with the organisation can sell or produce for themselves. This independence for producers was emphasised as crucial by stallholders in the market.

As such, the organisations of CEDEPO and APF are connected through local, national and international networks. CEDEPO and APF participate in the Mesa Provincial de Organizaciones de Productores Familiares de Buenos Aires (Provincial Bureau of Family Producer Organisations of Buenos Aires). Their aim is to strengthen grassroots organisations of peasants and family producers in the context of the 2001 crisis. Both groups are involved in an international movement MAELA (Movimiento Agroecológico de Latinoamérica y el Caribe or Agroecological Movement of Latin America and the Caribbean) and through MAELA are part of La Via Campesina, a global alliance of organisations of family farmers, peasant farmers, indigenous people, landless peasants and farm workers, rural women, and rural youth, representing at least 200 million families worldwide (Rosset and Martínez Torres, 2013). These examples demonstrate the multiple scales of organisation on an everyday life level and social movement sphere.

Mercado Bonpland is crucial for CEDEPO and APF, as the market provides a space where these small producers can sell without an intermediary. This gives local producers more control over what and when they sell, as well as an opportunity to sell at a fair price. The market therefore allows local co-operatives and communities to focus on improving their daily lives rather than solely focusing on production as in order to sell. Knowing that there is a secured sale through Bonpland therefore gives the producers more flexibility in focusing on production and to focus on improving autogestion.

The above examples demonstrate how Mercado Bonpland connects the supposedly separate yet related worlds of the urban and rural, production and consumption, nature and society. As with all markets they are reliant on production of surrounding regions to sell and vice versa. The market challenges linear relationships of production and consumption, reformulating organisation in order to create relationships of support. As such the space of Mercado Bonpland can only continue because of the many different spaces of autogestion being organised beyond Mercado Bonpland. Similarly, these autogestive workplaces, coops, farms and factories can continue to develop because of the work and space to sell provided by Mercado Bonpland.

In developing these reciprocal relationships, Mercado Bonpland (as well as the related examples of CEDEPO, APF, Movimiento Nacional Campesino e Indígena, La Vía Campesina, and other grassroots organisations), demonstrates that much of the work of such a market is in connecting with producers and organising in the neighbourhoods where production is situated, as much as in the market itself. As such, the market is entirely reliant on grassroots organising, much of which stems from the 2001 social movement scene. Each organisation in Bonpland also has to continue to work to connect with specific producers in small towns and through networks of producers. In this way, through producers, the connections between the way that products are produced, the labour inherent in producing them, the spatial configuration of this production and the sale on the marketplace of Bonpland, demonstrates alternative organisation processes. The interconnection of such networks, people, organisation and ideas is essential.

Mercado Bonpland's origins are in the experience of the crisis of 2001, and collectively organising to resist industrial agriculture in the case of the producers and to reclaim public spaces for the neighbourhood in the case of consumers. Mercado Bonpland, as an experiment in producing another economy, does not only have connections with producers for produce, it also develops relationships of support, knowledge exchange, visits and education projects. In addition, those that do not have a production relationship can use the market as networks of resources to expand their own potential to produce. This means that the expansion of networks of autogestive projects developed through organising together creates more possibilities, resources and connections between projects. Finally, these spaces are connected through people's use of Bonpland for their consumption, as they all support the development of healthy food, produced following agroecological principles. For Bonpland, the history and experience of organising developed these projects, and connects these sometimes seemingly disparate organisations.

Conclusions

In conclusion, we argue that Bonpland is one expression amongst many of an alternative and solidarity economy in practice. Constructed through networks of producers, this market relies on many different alternative spaces of production (such as family agriculture, reclaimed factories and cooperatives, and with the specific cases in this chapter: Titrayjú, Colectivo Solidario, La Asamblearia, CEDEPO and APF) that operate within and beyond the city. These relationships cannot operate without the support of the government and they are beyond utilitarian production interactions, relying on networks of organisers supporting multiple projects and exchanging practical and theoretical knowledge, workshops and events. Therefore, the networks of alternative economies that were formed from the experience of crisis of 2001 show connections and history of organising in this crisis period and beyond. Through these networks, constituted by neighbourhood production organising in Bonpland provides (re)claiming of unused spaces (market or land) for the local community and solidarity economy. As such we argue that seen through networks, Bonpland demonstrates challenges to the dominant understandings of the economy through production processes.

Bonpland organises urban and rural production through various grassroots social movements seeking to contest neoliberal production systems. On the one hand, the urban scale was arranged through assemblies then emerged from the crisis of 2001. On the other hand, rural movements are connected to Bonpland through the rural, peasant and indigenous movements (which have their origins in the 1960s and 1970s in Argentina) and more recent organisations of 'Family Agriculture'. These connections are strengthened by the changing perceptions of global environmental issues, which since the 60s, have resulted in farmers organising to produce in some cases in opposition of industrial agriculture. Additionally, urban consumers aware of these environmental issues wish to support alternative production through the consumption of ecological produce. The Bonpland experience generates understandings and practices that engage in consumption as a political act and that demonstrates the social relationships which exist in a market. The assembly of networks of different rural and urban social groups embodied in Mercado Bonpland actively creates a public space that facilitates this economy and demonstrates the recovery of the public space by a neighbourhood assembly. The physical space of the market demonstrates a fight for the occupation of public space in a context where there is a growing contestation over it.

Note

1　A kind of Sunday market that brings traditional Gaucho (Argentine cowboy) rural culture directly into the city.

References

Caracciolo, M. (2014). Economía Social y Solidaria: mercados y valor agregado en actividades rurales y urbanas. In: Rofman, A. and García, A. (Eds.), *Economía solidaria y cuestión regional en Argentina a principios de siglo XXI. Entre procesos de subordinación y prácticas alternativas.* Buenos Aires: Ediciones CEUR, 214–240. Available from: http://www.ceur-conicet.gov.ar

Carballo, C. (2000). *Las ferias francas de Misiones. Actores y desaíos de un proceso de desarrollo local.* Working document no 9. Buenos Aires: Centro de Estudios y Promoción Agraria.

Carballo, C., Tsakoumagkos, P., Gras, C., Rossi, C., Plano, J.L., and Bramuglia, G. (2004). Articulación de los pequeños productores con el mercado: limitantes y propuestas para superarlas. *Serie de Estudios e Investigaciones*, 7. Available from: http://www.ucar.gob.ar/index.php/biblioteca-multimedia/buscar-publicaciones/24-documentos/265-articulacion-de-los-pequenos-productores-con-el-mercado-limitantes-y-propuestas-para-superarlas

Coraggio, J.L. (2002). *La propuesta de economía solidaria frente a la economía neoliberal.* Conferencia sobre Economía Solidaria, Foro Social Mundial, Porto Alegre, Brasil. Available from: http://www.coraggioeconomia.org/jlc/archivos%20para%20descargar/ CHARLAS,%20CONFERENCIAS,%20DISCURSOS/JLC%20-%20Foro%20Social %20Mundial.pdf

Defourny, J. (2003). La autofinanciación de las cooperativas de trabajadores y la teoría económica. In: Chaves Ávila, R., Morales Gutiérrez, A.M. and Monzón Campos, J.L. (Coords), *Análisis económico de la empresa autogestionada.* Valencia: Centro Internacional de Investigación e Información sobre la Economía Pública, Social y Cooperativa, CIRIEC-España, pp. 183–212.

Fernández, A.M. (2011). *Política y subjetividad. Asambleas barriales y fábricas recuperadas.* Buenos Aires: Biblos.

García Guerreiro, L. (2010). Espacios de articulación, redes autogestivas e intercambios alternativos en la ciudad de Buenos Aires. *Otra Economía*, IV, 68–82.

Goldberg, C. (1999). *El Movimiento Agrario de Misiones en un escenario de transformación.* Tesis de Ingeniería Agronómica, Facultad de Agronomía, Universidad de Buenos Aires.

Herzer, H., Di Virgilio, M. and Rodríguez, C. (2015). Gentrification in the city of Buenos Aires: Global trends and local features. In: Lees, L., Bang Shin, H. and Lopez-Morales, E. (Eds.), *Global Gentrifications.* Bristol: Policy Press, pp. 199–222.

Hintze, S. (2007a). Políticas sociales argentinas 1990–2006. In: Vuotto, M. (Coord.), *La co-constitución de políticas públicas en el campo de la economía social.* Buenos Aires: Prometeo, pp. 3–23.

Hintze, S. (2007b). *Políticas sociales argentinas en el cambio de siglo: conjeturas sobre lo posible.* Buenos Aires: Espacio Editorial.

Laville, J. L., Lévesque, B. and Mendell, M. (2007). The social economy: Diverse approaches and practices in Europe and Canada. In: *The Social Economy: Building Inclusive Economies.* Paris: OECD, pp. 155–187.

La Asamblearia (2013). Cooperativa La Asamblearia: Mercado Bonpland [online]. Available from: http://asamblearia.blogspot.co.uk/p/mercadobonpland.htm (accessed 4 March 2014).

La Asamblearia (n.d.). Website. Accessed 1 May 2016. Available from: http://asamblea ria.blogspot.com.ar

Mauro, S. and Rossi, F. (2013). The movement of popular and neighborhood assemblies in the city of Buenos Aires, 2002–2011. *Latin American Perspectives*, 42(2), 107–124.

Medina, X. and Álvarez, M. (2009). El lugar por donde pasa la vida ... Los mercados y las demandas urbanas contemporáneas: Barcelona y Buenos Aires. *Estudios del hombre*, 24, 183–201.

Rosset, P. and Martínez Torres, M. (2013). Rural social movements and agroecology: Context, theory, and process. *Ecology and Society*, 17(3). Available from: http:// www.ecologyandsociety.org/vol17/iss3/art17/

Sevilla Guzmán, E. and Martínez Alier, J. (2006). New rural social movements and agroecology. In: Cloke, P.Marsden, T. and Mooney, P. (Eds.), *The Handbook of Rural Studies.* New York: SAGE Publications Ltd, pp. 468–479.

Singer, M.M. (2010). Who says "It's the economy"? Cross-national and cross-individual variation in the salience of economic performance. *Comparative Political Studies*, 44(3), 284–312.

Vázquez, L.J. (2006). *Evolución del Comercio Justo en Argentina: El Caso de la Yerba Mate Titrayju.* Tesis de Ingeniería Agronómica, Facultad de Agronomía de la Universidad de Buenos Aires.

Teubal, M. and Rodríguez, J. (2002). Globalización y sistemas agroalimentarios en la Argentina. In: Teubal, M. and Rodríguez, J. (Eds.), *Agro y alimentos en la globalización. Una perspectiva crítica.* Buenos Aires: La Colmena, pp. 63–94.

9 Public markets: Spaces for sociability under threat?

The case of Leeds' Kirkgate Market

Penny Rivlin and Sara González

Introduction

Traditional retail markets can function as spaces of sociability and inclusion for diverse groups in the city, yet these benefits are threatened by the advancement of urban neoliberal policies. In this chapter, we focus on Leeds' Kirkgate Market in the north of England, which after decades of local authority neglect is currently undergoing a process of transformation and redevelopment. An important historical and heritage asset to the city of Leeds, Kirkgate Market is one of the largest traditional markets in the UK and Europe, comprising both indoor and outdoor areas. Originating as a collection of open livestock, corn and street markets in the early 1800s, it was first constructed as a covered market in 1857, being subsequently adapted and extended (Fraser, 1980). In December 1975 a disastrous fire destroyed two thirds of the market, but left an extension built in 1904 unaffected. Subsequently, the market was classed as protected heritage, and two contiguous 'hangar structure' halls were constructed in 1976 and 1981 to replace the damaged sections.

Owned and managed by Leeds City Council (hereafter LCC), Kirkgate Market hosts around 400 businesses with various tenancy agreements, from casual day to long-term leases, that support over 2,000 jobs (LCC, 2010). Attracting an annual footfall of 10 million visitors (ibid.), the market offers an exceptionally diverse range of stalls; traditional greengrocers, butchers, fishmongers, bakers, florists and confectioners trade alongside drapery, haberdashery, clothing, DIY, homeware, and technology stalls. A substantial proportion of the goods and services reflect the ethnic and cultural diversity of market users, such that the scale and diversity of Kirkgate Market's offer contributes to its reputation as the most varied, affordably priced retail outlet in Leeds.

The market is situated in the heart of the city of Leeds, a city with an increasingly diverse population of 751,485 people, of which 18.9 per cent of residents identify themselves as Black and Minority Ethnic (BME; ONS, 2011). Considered to have a diverse and growing economic base, Leeds has been relatively resilient in the wake of the 2008 global financial crisis;

however, the city is spatially and socially segregated, with a significant proportion of residents living in areas of high deprivation (González and Oosterlynck, 2014). Exploring the significance of Kirkgate Market in this context, we will go on to empirically examine its role as an important space for affordable food provisioning and social inclusion, especially for residents from deprived neighbourhoods. Our interviews with market traders emphasise their role in servicing Leeds' diverse working class, low-income and marginalised communities, revealing a narrative of historical continuity affirmed in histories of British markets.

As explained in the introduction of this book, enclosed market structures were originally designed as a response to the unsanitary, disorderly and unmanaged open and street markets, in which livestock, agricultural produce, food and people circulated (Schmiechen and Carls, 1999). Late Georgian and Victorian enclosed markets were utilised by both the middle and working classes. However, the accessibility and affordability of perishable food ensured the continued patronage of the working classes throughout the nineteenth century (ibid, p. 21), such that for the urban poor, markets performed 'a function as important to the operation of urban life as the gas and waterworks' (Hodson, 1998, p. 98). In our research we have seen the continuity of this relationship between traditional markets and poorer communities supported in recent LCC data on Kirkgate Market: the majority of market users occupy the lower socio-economic class groups with students, female, middle-aged and elderly users predominating (LCC, 2014b). Attesting to the importance of the market for Leeds' BME and migrant communities, the dominant demographic profile of traders is female, aged 45 and over from BME backgrounds; it is worth noting that these communities are also disproportionately represented in the poorest residential wards of Leeds (LCC, 2011).

This chapter therefore situates Kirkgate Market as a key historical and societal landmark in the city of Leeds; it is part of the working class collective imaginary of the city, and more recently, a space of interaction for migrants and residents from ethnic minorities. Tracing developments over recent years, we argue that for Leeds' most vulnerable residents, this space of sociability is being transformed and potentially undermined through various local, national and global trends. We develop our argument as follows: First, we map out relevant theoretical concepts that were foregrounded in the introduction of this book. Second, we chart the recent history of Kirkgate Market within the context of austerity and the neoliberalisation of urban policy in Leeds and beyond. Third, we summarise Kirkgate Market as a space for sociability. Finally, the concluding section suggests that the benefits of sociability are threatened by the market's recent redevelopment and associated processes of gentrification.

The chapter utilises data from several sources. One of the authors has a long-term engagement as an action researcher in the market (see González and Waley, 2013), and more recently, we conducted ethnographic research as part of an interdisciplinary project.[1] We used multiple methods of investigation, including multimodal data collection, participant observation, and semi-structured interviews with market users; this chapter focuses on data generated from observation fieldnotes, narrative interviews and policy analysis (in a comparable way to González and Waley, 2013; Cattell et al., 2008; Watson and Studdert, 2006). We conducted semi-structured interviews with 9 female and 12 male traders aged between 28 and 64. In terms of ethnic origin, four traders identified as White British, three as White Eastern European, eight as British Asian, two as Asian, one as Chinese, one as Kurdish Syrian and two as Afghanistani. We identified 14 different 'main' languages (from those listed in ONS, 2011) spoken by traders, and proficiency in 35 other languages, all of which were deployed in the market. While we acknowledge that the study cannot claim empirical generalisation, we made concerted efforts, as far as possible, to ensure representativeness.

Markets as spaces for sociability and community diversity

There is growing evidence within the international literature that traditional retail markets represent a crucial node for the study of sociality in cities, and more broadly, for examinations of the economy–society nexus (Hiebert et al., 2015; Morales, 2011; Pottie-Sherman, 2011). Theorised from a range of per-spectives, the relationships between sociality and markets appear contingent on the specificities of place, spatial variation, historical (dis)continuity, cultural and economic practices and adaptations to these interconnecting processes. Historically, urban markets are key public spaces for the reception and integra-tion of strangers, new residents and immigrants to the city, facilitating com-munication across national, ethnic, linguistic, religious and socio-economic differences (Vicdan and Firat, 2015; Wessendorf, 2014; Morales, 2011). Con-trasting with public spaces such as neighbourhood parks, city squares and quasi-public spaces such as shopping malls, the comparatively unmediated, informal space of the city market can foster relations of recognition, adaptation and civility across multiple differences (Anderson, 2011).

Markets have been debated as spaces in which diversely-situated people actively 'create a public space' (de la Pradelle, 2006), where differences are mediated by 'rubbing along' (Watson, 2009) for the constitution of 'com-monplace diversity' in the city (Wessendorf, 2014). Everyday practices of mingling and observing in markets (Cattell et al., 2008) have fostered conditions whereby people conduct 'folk ethnographies' that may promote a 'cosmopolitan canopy' of social inclusion, cross- and inter-cultural under-standing, tolerance and civility (Anderson, 2011). A more ambivalent thread in sociologies of markets theorises social interaction across difference in terms of 'mutual avoidance' (Smith, 1965); a tacit 'indifference to difference' that can co-exist with 'openness to otherness' (Pardy, 2005); or the active assertion of difference and 'othering' to shore up power and position in an intensely competitive environment (Busch, 2010, and Liu, 2010 cited in Pottie-Sherman, 2011). This view interprets the market as a site of economic exchange relations wherein sociality is secondary or incidental to its primary function.

Crosscutting these contradictory tensions, another strand of research emphasises the role that markets play in mitigating social exclusion and economic disenfranchisement for marginalised groups. In their ethnography of eight UK markets, Watson and Studdert (2006) document the importance of markets as key spaces of social interaction, particularly for low-income groups. Drawing attention to class and gender dynamics in areas of socio-economic deprivation, the authors stress the significance of markets for the elderly, women and single parents with children, recommending their inclusion in national social exclusion policy agendas. Several studies on retail and street markets in London highlight a high correlation between areas of deprivation, high density BME populations and concentrations of street markets (Cross River Partnership, 2014; Hall, 2015; NEF, 2005), and their importance for migrant and local entrepreneurship, employment and social interaction (Dines, 2007; Hall, 2015; Morales, 2009); factors also highlighted in the chapter by González and Dawson in this book. In the context of the US, Morales' historical-empirical research of Chicago's Maxwell Street market (2009, 2011) argues that, historically, markets have functioned as social levellers, promoting economic and socio-cultural accommodation and assimilation in public space. The market's low barriers to entry and relatively informal, flexible and transparent regulatory system work to ameliorate discriminatory practices in employment and business sectors, affording viable pathways to social mobility to unskilled or undercapitalised people.

This body of evidence on the sociability and community benefits of markets for vulnerable and marginalised groups highlights their salience to cities within the context of urban diversity and socio-spatial inequalities associated with modernisation and urban redevelopment. Although our approach concurs with this perspective, our contention in this chapter is that the scaffold upon which these benefits are constructed and reproduced in UK markets is unstable and fragile. In part, this is due to the unique elements that constitute the market itself. The modes of inclusive sociability in public space that interest us here typically thrive in neglected, informal and under-the-radar places that have historically received limited attention from public authorities and private investors. But periods of neglect and abandonment in cities are often followed by phases of reinvestment, wherein formerly unattended spaces come under the spotlight of state and private interests (Smith, 1996). Ensuing urban redevelopment projects often trigger contested processes of transformation (e.g. Porter and Shaw, 2013; Rankin and McLean, 2015), bringing city markets under the remit of economic development agendas that generally prioritise commercial and economic interests (González and Waley, 2013). As the introduction and several other chapters of this book have discussed, these redevelopment agendas may exert pressure on markets to change through processes of increased regulation, formal governance and the sanitisation of space, creating conditions that can compromise provision of social and community benefits and contribute to the displacement of vulnerable market users and traders. In our contribution, we want to pursue the point that gentrification processes can compromise urban diversity and the associated sociability and community benefits that are realised in the unique space of the market.

Urban diversity and sociability in Leeds' Kirkgate Market: Accommodating difference in public space

This section empirically examines relations of sociability and urban diversity in Kirkgate Market. Our interviews with Kirkgate traders suggest that the market provides a unique space in Leeds for social interaction across a number of relational registers. In what follows, we discuss several interrelated dimensions of sociality that contribute to a conception of sociality as a valued resource that is a fundamental, rather than incidental, dimension of the market.

Kirkgate Market is a spectacle of cultural difference, encapsulating public multiculture not reproduced elsewhere in the city centre. Contrasting with the bland homogeneity of the nearby Trinity shopping mall and the stylised luxury of neighbouring Victoria Gate arcade (see Figure 9.2), the market's dominant aesthetic is of bustling eclecticism; on busy days the market assumes a vibrant, carnivalesque ambience of orderly disorderliness (Bakhtin, 1984). The socio-cultural and ethnically diverse backgrounds of market users and traders shape the range of inexpensive food, goods and services on offer – many of which are found exclusively in the market. Over several years of research engagement in the market, one of the authors has witnessed the proliferation of retail outlets such as mobile phone products and services, European, Asian and African-Caribbean styled hair and nails salons, clothing alteration stalls and ethnically-specific food and clothing outlets. Echoing other UK based studies of superdiverse city spaces (Hall, 2015; Wessendorf, 2014), these retail trends attest to Leeds' increasingly ethnically diverse and fluid migrant population (LCC, 2011).

Resonating with Wessendorf's (2014) conception of 'commonplace diversity' as normative in superdiverse neighbourhoods, all the traders we interviewed described the market as a 'mixed' environment that includes 'people from every background' and 'every community', encompassing 'all nationalities and languages'. Several longstanding British traders (of White and Asian origin) expressed evident pride in the diversity of market users, perceiving diversity as fundamental to its identity in the context of an increasingly homogenised and 'faceless' retail landscape:

The market's for everybody in Leeds; that's why it's different [...] We get people from all over the world coming in here, speaking different languages, from different cultures. If you just took a snapshot of the shop now, we've got a Chinese woman, we've got a Japanese woman at the back, an English woman; you know, just in that one second you've got maybe four different cultures there. You just don't get that in the high street.

(Male butcher, White British)

Another trader framed the market as a place of cultural familiarisation and integration for new migrants, explaining that 'all our customers from other countries are used to shopping in a market because most only have markets where they come from, so they prefer a market to a supermarket'. Notwithstanding these positive responses to diversity, we were mindful that the traders' recognition of diversity as a unique dimension of the market does not necessarily evidence active social engagement and connection across difference (Pardy, 2005). Like other studies of sociality in markets (e.g. Anderson, 2011; Mele et al., 2015; Watson, 2009; Watson and Studdert, 2006) we observed spontaneous enactments of sociability between traders and passers-by such as smiles, nods and waves between traders and customers of diverse backgrounds – the kind of minimal engagement described by Watson (2009) as 'rubbing along' that constitutes a distinctive 'social glue' in markets. However, our research data revealed that as a means of accommodating the social mix of the market, traders engaged in a fluid process of adaptation and negotiation, which in turn fostered transcultural social connections.

Strategies of accommodation originated in economic concerns; the traders were reflexive about 'keeping pace' with the needs of an increasingly diversified customer demographic. As noted earlier, ethno-cultural accommodation is explicitly evidenced by the growth of stalls, providing for a fluid market multiculture. We interviewed several recent and longer-established migrant traders hosting cafés or fast-food outlets serving Asian, Halal and Middle-Eastern food, and other stalls offering goods and services familiar to specific ethnic, faith and national backgrounds. These traders explained that their stalls provided unique opportunities – particularly to those at early stages of migration – to maintain attachments to the cultural and communicative repertoires, traditions and preferences of their countries of origin, thus contributing to a sense of 'home' in a new and unfamiliar environment.

Yet we also observed practices of adaptation and accommodation in the market's long-standing and traditionally 'British' stalls. Responding to the shifting ethno-cultural market landscape, traders were engaged in a sustained process of intercultural learning, diversifying their product lines accordingly. A White British trader in beauty products pointed out her expanding range of 'different skin tone' cosmetics and 'alcohol-free' fragrance options sourced for her 'Muslim and Asian customers', many of whom would visit her stall 'just to chat', despite the episodic nature of beauty consumption. Similarly one butcher commented that the wide range of offal and other cuts on display now constituted between 40 and 50 per cent of sales, effecting a shift from a predominantly White British customer base:

Take pig tongues; a perfect example. You would never see them in a supermarket, you would probably never see them in a normal butcher's shop on the high street. We purposely bought those in for our Chinese, our Asian and African customers.

(Male trader, White British)

In addition to the more instrumental practices of product diversification, traders employed a range of communicative repertoires that contributed to the sociability of the market. We observed several traders exhibiting cultural competency via the appropriation of material cultural signifiers. The butcher quoted above explained the display of a Chinese lucky cat in the shop window as 'a wave "hello" to our Chinese customers', noting that 'the kids love that, too' (see Figure 9.1). Similarly, a fishmonger exhibited cultural competency in relation to his Chinese customers via the display of celebratory Chinese New Year banners and red paper lanterns. He also told us that he had learned to practice culturally-specific modes of fish preparation to accommodate the beliefs and food habits of customers from different ethno-cultural and national backgrounds.

A striking feature of market users' social interactions was the range of strategies employed to bridge language differences and proficiencies. Although

Figure 9.1 Chinese lucky cat displayed in butcher's shop at Kirkgate Market
Photo: Penny Rivlin, 2015.

traders' and customers' English language skills ranged from fairly basic to fluent, language proficiency did not appear to constrain social interaction more generally, or specifically, forms of inter-cultural and inter-ethnic exchange. In particular, an emphasis on the practice of patience was well-rehearsed in traders' narratives, suggesting that socio-cultural diversity is re-shaping the rhythms and temporalities of consumption and sociality in the market along 'slow' lines (Paiva et al., 2017). Traders typically kept notepads and pens for writing the prices of goods, or to communicate through simple drawings. In parti-cular, mobile technologies played an important role in mediating language differences in the market. Many traders reported that customers routinely deployed their mobile phones as translation devices via the use of language translation applications and dictionaries, online images of the desired product, or for engaging an English-speaking friend or family member as translator. These communicative strategies were complemented with informal linguistic interactions that involved traders' and customers' acquisition of basic words and greetings in others' main languages besides English: as one trader commented, 'we learn from them, and they learn from us'.

Kirkgate Market therefore combines spaces for specific ethnic minorities who are attracted by speciality produce, as well as more general stalls where traders and users from different backgrounds and cultures accommodate each other through various communication strategies. These reciprocal practices of accommodation turn the market into a 'comfortable' space of social and cultural inclusion, which, for many traders, is key to its uniqueness within the Leeds' retail landscape. Here, it is worthwhile recounting the narratives of two recent migrants to Leeds. One young Polish man related his initial concerns that his English language skills 'would not be enough' to forge and sustain mean-ingful relationships within the market, going on to explain that he quickly felt part of a 'community' of market users:

> I am very comfortable here; my English – it's been no problem at all in this community [...] and I try to make people feel comfortable. For example, some Muslim women wearing the veil are unable to speak English. It needs a certain amount of social skill [...] because they might be embarrassed that they don't know English. And for them to still feel comfortable, I do a bit of pretending I don't need their English.
>
> (Male trader, Polish food shop)

Another young trader's narrative of comfort suggests that the market opens up possibilities for local, elective affinities to space and place (cf. Savage et al., 2005) for diasporic subjects:

> Nobody discriminates here. I am comfortable here [...] people are friendly in the market; the customers, the traders. I think the market is my com-munity now.
>
> (Confectionery trader, Afghani origin)

Crucially, some traders' narratives of the market as a site of comfort and elective affinities were explicitly framed in relation to social class differences, particularly in relation to low-income users. One British Asian trader of 30 years standing observed that ethnic and linguistic diversity in the market is as 'commonplace' (Wessendorf, 2014) as the class position of its primary users:

> It doesn't matter whether the customers are White, Black, Asian, Polish or whatever – they're nearly all working class in the market.
>
> (Male, British Asian, homewares trader)

Allied to this recognition of the market's function as an inclusive space for low-income users was a concern for the welfare of the elderly population in the context of the spatial changes to the market, and the council's corresponding decision to remove casual seating to deter homeless people and anti-social behaviour:

> People come to the market because they're lonely, to pass the time; especially the elderly come and walk around. And there's nowhere really for them to sit unless you go to the cafés, which costs them money, and they probably can't afford it, you know, to spend money. There should be more places for the elderly to sit.
>
> (Male trader, South Asian origin)

Contesting management policy via an ethic of care, this trader positioned two chairs adjacent to his stall for the benefit of elderly market users – a well-used facility that contributed to the overall inclusive sociability of the market. We observed traders institute other everyday acts of care, ranging from the storage of customers' shopping bags, to providing hot drinks for regular elderly and low-income users. On a day of heavy rain, we observed a trader leave his stall to drive an elderly woman home after noticing her soaked appearance, another trader stepping in to watch his stall in his absence. That the prosaic encounters described throughout this section require patient, repeated investments in time, effort and care suggests that market users actively seek and value inclusive sociality in ways that disrupt conceptualisations that foreground markets as sites of indifference to, or avoidance of 'otherness', wherein economic exchange relations predominate.

Gentrification pressures and threats to inclusive sociability

As we stated in the introduction, this chapter discusses the importance of Kirkgate Market as a space of sociability across diverse relational registers, but questions the extent to which this sociability will be maintained in the face of recent transformations in the market led by the local authority, Leeds City Council (LCC). To understand these transformations, this section analyses the recent history of LCC's approach to the market, considering its selective process of policy development, as well as wider urban transformations. Informed by Hackworth's (2002, p. 815) distillation of gentrification processes as 'the production of urban space for progressively more affluent users', we argue that LCC's redevelopment of the market is a representative example.

Up until the 1970s, Kirkgate Market was an important reference point for the population of Leeds as both a provisioning space and part of the local identity – in particular for working class communities and other marginalised groups. Since then, however, an array of interrelated processes led to a slow phase of decline. Significantly, as mentioned earlier, a fire destroyed about two thirds of the market in 1975, which led to the construction of two utilitarian buildings of lower quality (Burt and Grady, 1992), and since then LCC has struggled to institute plans to support and maintain the infrastructure of the market. In the mid-1980s a planning proposal for a megaproject advanced by a private Dutch company would have resulted in much of the market being demolished and its marginalisation within a modern shopping centre. This proposal encountered substantial opposition from traders, members of the public, civic organisations and public campaign support, totalling 250,000 signatories, eventually resulting in its abandonment following a public inquiry (Lytton, 1988). Following this defeat, LCC undertook a series of improvements in the early 1990s, especially to the heritage-listed front profile of the market. However, two further decades of minimal investment in infrastructural maintenance, unaddressed until the end of the 2000s, has resulted in a general condition of neglect.[2]

This policy and (dis)investment history has deteriorated the relationship between traders and LCC, and more generally, has tainted the public's opinion of LCC's intentions towards the market. Indeed, one butcher who is currently defying LCC's recent relocation policy to a newly designated area has displayed a newspaper clipping from the 1986 redevelopment episode, telling us that 'history was repeating itself'. This process of neglect of the public market can be explained by a more general shift in the role and function of local authorities throughout the UK, and more widely throughout the western world. Since the rise of neoliberal ideology and policy from the 1970s, the local state is reconceptualised as an entrepreneurial economic actor that actively seeks investment and income generation opportunities, signalling a shift from its primary function in the delivery and distribution of public service (Harvey, 1989). Local authorities across the UK have thus marginalised public markets and no longer tend to see them as part of their welfare provision; instead they have treated them as 'cash cows' (House of Commons, 2009), or increasingly, have opted to sell them to private developers under the remit of a wider austerity agenda (González and Dawson, 2015).

In parallel with the abandonment of the market as a public service, LCC pursued a policy of regeneration of the city centre from the late 1990s into the 2000s that aligned with a broader national policy of 'urban renaissance' (Imrie and Raco, 2003). In Leeds, this renaissance took the form of a huge expansion of residential use in the city centre through the building of 9,500 apartments, primarily constructed between 2003 and 2009 (Unsworth, 2010). This corresponded with an upgrade of the core shopping district, most recently advanced by the construction of the new Trinity shopping centre (see Figure 9.2). Additionally, urban renaissance policy also fueled the development of smaller versions of the large supermarket chains, enabling their insertion into the existing fabric of town centres and direct competition with the publicly-owned market (see Figure 9.2 for new supermarkets in Leeds city centre). González and Hodkinson (2014) analyse this renaissance as a process of unfinished new-build gentrification, which, whilst not directly displacing residents, has created new exclusionary residential and commercial urban landscapes unaffordable for many in the city and city-region.

Part of this renaissance incorporated the area immediately adjacent to Kirkgate Market (see Figure 9.2), a similarly neglected site formerly occupied by municipally-owned car parks, which had been earmarked by LCC for years as a retail development opportunity (LCC, 2005). Protracted negotiations with private developers eventually led to the construction of an additional £150m luxury shopping centre, 'Victoria Gate', which opened in October 2016. Anticipating potential opportunities for Kirkgate Market in relation to the Victoria Gate development, LCC eventually launched an independent redevelopment plan for the market in 2012, supported by borrowing £13m based on future profits. Completed in 2016, this public-led redevelopment project has important elements of physical redevelopment in conjunction with aims to change the public image of the market through a new a branding

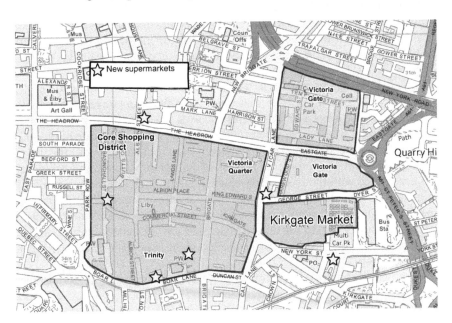

Figure 9.2 Map of Leeds city centre
Source: Annotated Ordnance Survey map by the authors.

strategy. The redeveloped market now hosts a revamped 'fresh food' area that showcases the heritage aspects of the building – for the first time bringing butchers and fishmongers together. It also cleared one of the Halls (evicting many traders) built after the 1975 fire to facilitate a large 'meet and eat' space featuring nine street food cafés. Other works included partial repair to the roof and the improvement of waste management and electrical facilities.

Whilst this capital investment is long overdue, our contention is that the redevelopment of Kirkgate Market reflects aspirations aimed at capturing customers and traders from or oriented to a higher socio-economic demographic, establishing a trend of gentrification. In the final planning stages in 2014, such intentions were implied in the market manager's address to local press: 'If we are going to be taken seriously as a retail presence in the centre of Leeds, we have to be accessible to our audience, whether that means targeting a younger, student audience or the middle class shopper' (Hudson, 2014). The appeal to 'be taken seriously' as a city centre retailer is predicated on the imagined participation of the middle-class consumer-citizen upon which neoliberal policy is based. At the same time displacement is inferred by what is *excluded* from the vision of a productive retail market: an existing, 'deficient' customer base constituting the elderly, new migrants, the unemployed and working class market users who collectively lack the material and cultural resources to be taken seriously (see also Rankin and McLean, 2015; Sullivan and Shaw, 2011). From the first phases of planning of this redevelopment project, LCC has expressed minimal concern over its potential social impact. Evidenced as early as 2010, a LCC strategy document questioned 'whether the market really is the best place for those on low incomes to shop [...] there are a wealth of alternative discount and supermarket offers located within the city centre and communities, many of which offer cheaper alternatives' (LCC, 2009, p. 20).

In their discussion of markets as the 'new frontiers' of gentrification, Gonzalez and Waley (2013) demonstrate that one of the first steps taken towards the replacement of traders and customers involves the selective displacement of long-established traders that do not 'fit' with the profile of upscaled branding. Whether strategically-deployed by market management or not, this trend corresponds with the long-term neglect of Kirkgate Market, and moreover, has accelerated with the recent redevelopment. Although it has proved difficult to qualify this trend, we do know that the redeveloped market has less trading space, and that many established traders were unable to secure suitable spaces in the new areas due to significant rent rises and associated costs – factors which implicate a process of trader displacement (Friends of Leeds Kirkgate Market, 2014b).

We observed simultaneous gentrifying initiatives in the realm of food. Earlier, we explained that Kirkgate Market has long functioned as a key space for everyday provisioning in the city centre, providing good quality, affordable food. This provisioning for instrumental needs particularly benefits Leeds' poorer residents, especially those living in disinvested locales and wards that suffer underprovision of accessible and affordable food retail (e.g. Wrigley et al., 2003). Although the price point in the market is comparatively low in relation to the city centre foodscape, some elements are exceptionally so. For instance, a number of fresh food stalls sell bowls of mixed fruit and vegetables at £1; others offer heavy discounts on packs of perishable food in the last hour of trading. However, recent initiatives in LCC's redevelopment agenda suggest that this important dimension of the market is at risk of marginalisation and even displacement. One prominent indicator includes LCC's recent trader recruitment drive, which clearly aims at refashioning the market's food sector along gourmet lines. The council's tender pack invites traders with expertise across a wide range of 'niche' foods including: World Food, Artisan Breads, Patisserie, Delicatessen, Smoothies, Handmade Chocolates, Organic and Local Farm products, Niche Supermarkets and Micro Brewery and Wine Retail (LCC, 2016). Other indicators of food gourmetisation include the installation of six international 'street food' inspired kiosks and specialist coffee bars in the new events space, and a project trialling 'pop-up' independent food and drink businesses. Signalling aims to attract a new socio-economic demographic, the council claims that these latter food ventures will provide 'something unique and exclusive to the market [...] encouraging new customers and traders to use the market', thus positioning 'Kirkgate Market as a supporter of high quality independent food and drink businesses' (LCC, 2015).

We can see in this rebranding of the food offer of Kirkgate Market the appropriation, sanitisation and commodification of ethnically and culturally diverse food cultures; they are re-packaged as 'unique and exclusive' gourmet experiences and 'othered' for the consumption of the potential customer base that the market management wishes to target. The problem here lies in the tendency for price elevation in line with professional incomes and therefore patronage (Anguelovski, 2015). This gourmetisation process can also lead to the displacement of long term market users, a point underscored by Sullivan and Shaw in their study of retail gentrification and class in the US: 'New retail offers goods and services that cater to newcomers, charge prices that correspond to professional incomes, and create cultural symbols and spaces that tend to attract newcomers but alienate long-time residents' (Sullivan and Shaw, 2011, p. 414).

The Leeds-based campaign group Friends of Leeds Kirkgate Market challenged this potentially exclusionary aspect of the market redevelopment project, objecting on the basis that it 'presents genuine threats to the Market's future survival as a community hub' (Friends of Leeds Kirkgate Market, 2014b). Delineating the ways in which the council had failed to study the customer base of the market, the group disputed its capacity to assess the social impact of the project. In response, LCC acknowledged that the 'community impact assessment' was insufficiently researched, resubmitting an alternative that nevertheless failed to address any of the objections raised by the campaign group (Friends of Leeds Kirkgate Market, 2014a).[3]

Similar concerns in relation to the gentrification of the market have been raised by traders, market users, and other members of the public at various stages of the planning process. Of particular note, a key LCC consultation document *Statement of Community Involvement*, commissioned in 2014, summarises the responses to the proposed redevelopment in an on-site and online public survey. Worth quoting at length, the following extract captures Leeds residents' anxieties in the early stages of gentrification, and their efforts to contest its further progression:

> It is clear is that the majority of customers, traders and visitors to the market enjoy the heritage feel of the market and this should be maintained. Some consultees felt that *the proposed designs are sterile and there is a concern that Kirkgate Market will be gentrified,* or the unique character and atmosphere of the building will be lost through the refurbishment. *Comments made indicated that Kirkgate Market is a place where people feel welcome regardless of their social status and that this should be maintained. Some felt that the only way this can happen is to keep rents low therefore keeping product prices low.* Comments on the online questionnaire raised concerns that as a result of the development the market may become gentrified and prices would increase.
>
> (LCC, 2014a, p. 21, italics in the original)

It is also the case, however, that some traders expect to benefit from, and would welcome the arrival of, different customer traffic generated from the newly-opened Victoria Gate shopping centre. Given the long period of neglect of the market by the council and recent disruptions associated with the redevelopment project, many traders have lost the patronage of long-term customers, and as such, look forward to the arrival of a more affluent clientele. However, at the present juncture it would be premature to evaluate the extent to which the market can balance the anticipated customer mix, nor how many of the current traders will realistically be able, or indeed willing, to adapt their products, service and modes of accommodation in response to a changing market environment.

Conclusions

In this chapter we have presented public markets as complex spaces of both value and tension. Markets serve as places for the reproduction of sociability and inclusion in increasingly diverse cities, but this valuable and valued resource is threatened by urban redevelopment processes. In pursuing this contradiction we have identified a range of interconnected risk elements that can advance the displacement of long-term market users and traders, pushing traditional retail markets in the direction of gentrification. We have discussed these complex issues in the case of a large public market in the centre of Leeds that has historically served a very diverse customer base, more recently welcoming new migrants from around the world as both traders and users. Through our ethnographic research we have found that the market functions as a space for dense sociability where market traders and users routinely engage in practices of cultural translation, adaptation and accommodation. In much of the literature on markets as places for social interaction, researchers have highlighted the powerful contribution of such semi-public spaces for the development of low level, non-threatening interaction between diverse publics, which in other spaces might be in conflict or competition due to class, culture and ethnic differences. However, our research reveals that traders and users do not merely tolerate each other to facilitate instrumental economic transactions; rather, they invest considerable time, care and effort in everyday interactions to accommodate difference and communication across cultural, class and language barriers.

We have also shown how Kirkgate Market is at a critical juncture of potential transformation in response to a major process of redevelopment that has involved a partial upgrading of the infrastructure and the relocation and subsequent displacement of many traders. We have argued that this redevelopment project is yoked to the opening of an adjacent luxury shopping centre from which market managers and LCC aim to attract a more affluent customer base to Kirkgate Market. We have demonstrated that this state-led redevelopment project has exhibited limited consideration for the locally-valued social and intercultural aspects of the market, therefore potentially posing a risk to the vibrant forms of inclusive sociability identified earlier in this chapter. This sociability is facilitated by a number of intersecting elements: affordability; familiarity and comfort at spatial and cultural levels; a particular mix of uses; and the socio-economic and ethnic mix of market users. We contend that the trend towards gentrification triggered by the redevelopment project is potentially eroding these unique socio-spatial elements, as well as the history of traditional retail markets as a public service. At the present moment, we speculate that this constellation of elements would be exceptionally difficult to reproduce both in the market itself, and elsewhere across Leeds' retail landscape and its city centre as a whole. Thus, as a fitting end, we borrow a biological metaphor from a London-based campaign group that likens the market to a 'human coral'; 'Rather like a coral reef this multi-layered and multi-faceted community is a fragile form, easily destroyed, yet near impossible to replicate' (Friends of Queen's Market, n.d.).

Notes

1 'Leeds Voices: Communicating Superdiversity in the Market' was funded by the British Academy and the University of Leeds between October 2015 and July 2016 (http://voices.leeds.ac.uk). Through the twin lens of superdiversity and multi-modality, the project explored the multilingual and cultural accommodations that people of different ethnicities make in their interactions at Kirkgate Market (Blommaert, 2014; Vertovec, 2007).

2 These data on investment were uncovered by public inquiries made by the campaign group Friends of Leeds Kirkgate Market, to which one of the authors belongs. They are available through the website whatdotheyknow.com.

3 The acknowledgement of the deficient Equality Impact Assessment came in private emails between Leeds City Council policy officers and members of the Friends of Leeds Kirkgate Market campaign. These are accessible upon request.

References

Anderson, E. (2011). *The Cosmopolitan Canopy: Race and Civility in Everyday Life.* New York: W.W. Norton & Company Inc.

Anguelovski, I. (2015). Alternative food provision conflicts in cities: Contesting food privilege, injustice, and whiteness in Jamaica Plain, Boston. *Geoforum*, 58, 184–194.

Bakhtin, M. (1984). *Rabelais and His World.* Bloomington, IN: Indiana University Press.

Blommaert, J. (2014). Commentary: Superdiversity, old and new. *Tilburg Papers in Culture Studies.* Paper 105. Available from: https://www.tilburguniversity.edu/uploa d/048195df-cf84-417b-8fc0-51078c6eda0a_TPCS_105_Blommaert.pdf

Burt, S. and Grady, K. (1992). *Kirkgate Market: An Illustrated History.* Published by Steven Burt and Kevin Grady.

Busch, D. (2010). Shopping in hospitality: Situational constructions of customer-vendor relationships among tourists at a bazaar on the German-Polish border. *Language and Intercultural Communication*, 10(1), 72–89.

Cattell, V., Dines, N., Gesler, W. and Curtis, S. (2008). Mingling, observing, and lingering: Everyday public spaces and their implications for well-being and social relations. *Health and Place*, 14(3), 544–561.

Cross River Partnership (2014). *Sustainable Urban Markets: An Action Plan for London.* London: Cross River Partnership. Available from: http://crossriverpartnership.org/ media/2014/12/Sustainable-Urban-Markets-An-Action-Plan-for-London3.pdf

Dines, N. (2007). The experience of diversity in an era of urban regeneration: The case of Queens Market, East London. EURODIV Working Paper 48. Milan: Fondazione. Eni Enrico Mattei. Available from: http://ebos.com.cy/susdiv/uploadfiles/ED2007-048.pdf

Fraser, D. (1980). *A History of Modern Leeds.* Manchester: Manchester University Press.

Friends of Leeds Kirkgate Market (2014a). FoLKM's speech to the Planning Committee, 12th December. Friends of Leeds Kirkgate Market blog. Available from: https://kirkga temarket.wordpress.com/2014/12/12/folkms-speech-to-the-planning-committee/

Friends of Leeds Kirkgate Market (2014b). Objection to Kirkgate Market Refurbishment, Works Planning Application (14/04516/LA). Available from: https:// kirkgatemarket.files.wordpress.com/2014/09/folkm-comments-to-14-04516-la-kirk gate-market.pdf

Friends of Queens Market (n.d.). Website. Available from: http://www.friendsofqueensma rket.org.uk/index2.html

González, S. and Dawson, G. (2015). Traditional markets under threat: Why it's happening and what can traders and campaigners do. Available from: http://tradmarke tresearch.weebly.com/uploads/4/5/6/7/45677825/traditional_markets_under_threat-_ full.pdf

González, S. and Hodkinson, S. (2014). Gentrificación como política pública en una ciudad provincial. El caso de la ciudad de Leeds en el Reino Unido. *Revista de Geografía Norte Grande*, 58, 93–109.

Gonzalez, S. and Oosterlynck, S. (2014). Crisis and resilience in a finance-led city: Effects of the global financial crisis in Leeds. *Urban Studies*, 51(15), 3164–3179.

González, S. and Waley, P. (2013). Traditional retail markets: The new gentrification frontier? *Antipode*, 45(4), 965–983.

Hackworth, J. (2002). Postrecession gentrification in New York city. *Urban Affairs Review*, 37(6), 815–843.

Hall, S. (2015). Super-diverse Street: A 'trans-ethnography' across migrant localities. *Ethnic and Racial Studies*, 38(1), 22–37.

Harvey, H. (1989). From managerialism to entrepreneurialism: The transformation in urban governance in late capitalism. *Geografiska Annaler*, 71B(1), 3–17.

Hiebert, D., Rath, J. and Vertovec, S. (2015). Urban markets and diversity: Towards a research agenda. *Ethnic and Racial Studies*, 38(1), 5–21.

Hodson, D. (1998). "The municipal store": Adaptation and development in the retail markets of nineteenth-century urban Lancashire. *Business History*, 40(4), 94–114.

House of Commons (HoC) (2009). Market failure? Can the traditional market survive? Ninth report of session 2008–2009. Communities and Local Government Committee. Available from: http://www.publications.parliament.uk/pa/cm200809/cm select/cmcomloc/308/308i.pdf

Hudson, N. (2014). Major revamp for Leeds' historic market. *Yorkshire Post*, 4 February 2014. Available from: http://www.yorkshirepost.co.uk/news/analysis/major-revamp-for-leeds-historic-market-1-6415329#ixzz47aydXxGW

Imrie, R. and Raco, M. (2003). *Urban Renaissance? New Labour, Community and Urban Policy*. Bristol: Policy Press.

LCC (Leeds City Council) (2005). Eastgate and Harewood: Supplementary planning document, October 2005. Available from: http://www.leeds.gov.uk/docs/FPI_EAH_001%20Eastegate%20and%20Harewood%20SPD.pdf

LCC (Leeds City Council) (2010). Towards a strategy for Kirkgate Market: The evidence base. 9 December 2010. Available from: http://democracy.leeds.gov.uk/m gConvert2PDF.aspx?ID=52088

LCC (Leeds City Council) (2011). Leeds – the big picture: A summary of the results of the 2011 Census. Available from: https://observatory.leeds.gov.uk/resource/view? resourceId=3759

LCC (Leeds City Council) (2014a). Kirkgate Market improvement and refurbishment scheme: Statement of community involvement. Available from: http://www.leeds. gov.uk/c/Documents/Kirkgate%20Market/Stetement%20of%20Community% 20Involvement.pdf

LCC (Leeds City Council) (2014b). Kirkgate Market demographic profiling – report. 12 November 2014. Available from: http://plandocs.leeds.gov.uk/WAM/doc/Ba ckGround%20Papers-1152274.pdf?extension=.pdf&id=1152274&location=Volum e4&contentType=application/pdf&pageCount=1

LCC (Leeds City Council) (2015). Procurement of a manager tenant for pop-up food units in Kirkgate covered daily market. Market testing exercise. Available from: http://www.leeds.gov.uk/leedsmarkets/News/Documents/Market%20Testing%20Ma nager%20Tenant.pdf

LCC (Leeds City Council) (2016). Trade in fresh produce: Commercial tender pack. Available from: http://www.leeds.gov.uk/leedsmarkets/Pages/Tender-pack-informa tion.aspx

Liu, J.J. (2010). Contact and identity: The experience of 'china goods' in a Ghanaian market-place. *Journal of Community and Applied Social Psychology*, 20, 184–201.

Lytton, K.G.B. (1988). Report to the Secretary of State for the Environment of a public inquiry held in April-May 1988 in connection with a Compulsory Purchase

Order (Kirkgate Market Area) made by Leeds City Council. Available from the authors.

Mele, C., Ng, M. and Bo Chim, M. (2015). Urban markets as a 'corrective' to advanced urbanism: The social space of wet markets in contemporary Singapore. *Urban Studies*, 52(1), 103–120.

Morales, A. (2009). Public markets as community development tools. *Journal of Planning Education and Research*, 28, 426–440.

Morales, A. (2011). Marketplaces: Prospects for social, economic, and political development. *Journal of Planning Literature*, 26(1), 3–17.

NEF (2005). *Trading Places: The Local and Economic Impact of Street Produce and Farmer's Markets*. London: NEF.

ONS (Office for National Statistics) (2011). *Statistical Bulletin: 2011 Census – Population and Household Estimates for England and Wales*. London: Office for National Statistics.

Paiva, D., Cachinho, H. and Barata-Salgueiro, T. (2017). The pace of life and temporal resources in a neighbourhood of an edge city. *Time & Society*, 26(1), 28–51.

Pardy, M. (2005). Kant comes to Footscray Mall: Thinking about local cosmopolitanism. In: Long, C., Shaw, K. and Merlot, C. (Eds.), *Sub Urban Fantasies*. Melbourne: Australian Scholarly Publishing, pp. 107–128.

Porter, L. and Shaw, K. (Eds.) (2013). *Whose Urban Renaissance? An International Comparison of Urban Regeneration Strategies*. London: Routledge.

Pottie-Sherman, Y. (2011). Markets and diversity: An Overview. Working Paper, 11–03. Göttingen: Max-Planck Institute.

de la Pradelle, M. (2006). *Market Day in Provence*. Chicago University Press, IL: Chicago.

Rankin, K.N. and McLean, H. (2015). Governing the commercial streets of the city: New terrains of disinvestment and gentrification in Toronto's inner suburbs. *Antipode*, 47(1), 216–239.

Savage, M., Bagnall, G. and Longhurst, B. (2005). *Globalization and Belonging*. London: Sage.

Schmiechen, J. and Carls, K. (1999). *The British Market Hall: A Social and Architectural History*. New Haven, CT and London: Yale University Press.

Smith, M.G. (1965). *The Plural Society in the British West Indies*. Berkeley, CA: University of California Press.

Smith, N. (1996). *The New Urban Frontier. Gentrification and the Revanchist City*. London: Routledge.

Sullivan, D.M. and Shaw, S. (2011). Retail gentrification and race: The case of Alberta Street in Portland, Oregon. *Urban Affairs Review*, 47(3), 413–432.

Unsworth, R. (2010). *City Living Beyond the Boom: Leeds Survey 2010*. A collaboration with five firms of agents in the city. A report by Knight Frank, Morgans City Living, Savills, King Sturge and Allsop. Available at: http://www.propertyweek. com/Journals/Builder_Group/Property_Week/05_March_2010/attachments/Leeds% 20City%20Living%202010%20report%20-%20Knight%20Frank,%20Morgans% 20City%20Living,%20King%20Sturge,%20Allsop,%20Savills,%20University%20 of%20Leeds.pdf

Vertovec, S. (2007). Super-diversity and its implications. *Ethnic and Racial Studies*, 30, 1024–1054.

Vicdan, H. and Firat, F. (2015). Evolving desire to experience the social 'other': Insights from the high-society bazaar. *Journal of Consumer Culture*, 15(2), 248–276.

Watson, S. (2009). The magic of the marketplace: Sociality in a neglected public space. *Urban Studies*, 46(8), 1577–1591.

Watson, S. and Studdert, D. (2006). *Markets as Sites for Social Interaction: Spaces of Diversity*. Bristol: The Policy Press and Joseph Rowntree Foundation.

Wessendorf, S. (2014). *Commonplace Diversity: Social Relations in a Super-diverse Context*. Basingstoke: Palgrave Macmillan.

Wrigley, N., Warm, D. and Margetts, B. (2003). Deprivation, diet, and food-retail access: Findings from the Leeds 'food deserts' study. *Environment and Planning A*, 35(1), 151–188.

10 Contested identities and ethnicities in the marketplace

Sofia's city centre between the East and the West of Europe[1]

Stoyanka Andreeva Eneva

Introduction

Popularly known as the Women's Market, the central market of Sofia was built at the end of the nineteenth century. There are different versions of the origins of this name; according to Kostentzeva (2008) it was one of the few public spaces that could be freely and autonomously accessed by women during the early twentieth century. Given the long and intense bargaining processes when buying, and the use of this space as a meeting point, women spent most of the day doing the grocery shopping.

Though the location of this market has been slightly modified, it can still be found in the central area of the city. The appearance of the place has also changed through the years, but it has maintained a certain level of informality and ability to exist with little administrative regulation. However, local authorities have attempted to eliminate these features through a series of interventions in 2006 and 2014. The first of these initiatives aimed at reorganising most of the market stalls through the installation of PVC roofing panels (see Figure 10.1). The measures implemented in 2014 adversely affected the layout of the space, since they eliminated all of the remaining open-air shops, which were replaced by the construction of five wooden buildings housing 90 kiosks intended for the provision of services and non-perishable food products.

These two interventions attempted to modernise the market; however, the second measure – apart from aiming to revitalise this area through the creation of new stores and restaurants – was marked by the 'Europeanisation' of heritage, a change in the commercial offer and, even more starkly, the tenants. The racialisation of poverty, illegal trading activities and crime were used to present this market and its neighbouring areas as a problematic and undesirable space.

Markets are often associated with cultural diversity, since they offer unique spaces for the convergence of diverse cultural groups (Watson, 2009). However, the emergence of exacerbated stereotypes, hostilities and ethnic and/or social boundaries (Thuen, 1999) can prevent this relationship from being fluid and equal. In this sense, this chapter – through the study of the central market of Sofia – aims to place ethnic identities as a key element of daily activities within the context of markets.

Figure 10.1 Part of the 'Women's Market' after it was reorganised in 2006
Source: Stoyanka Andreeva Eneva.

The opening of borders in Eastern Europe in the 1990s not only affected personal lives but also commercial routes such as the series of networks used by trade and people during the socialist era (Apostolova, 2014). Likewise, markets offered employment opportunities and national and social mobility for some ethnic minorities, thus becoming havens for socially- and spatially-excluded groups. In this sense, one of the main characteristics of the Women's Market during the 1990s was its association with concepts such as mobility and the possibility of finding affordable and rare goods that were difficult to find in regular stores. This was important in a period marked by shortage, inflation and low-paid employment. This situation led to the emergence of new stores located beyond the limits of the market: the opening of shops in the first floors of neighbouring buildings and the arrival of the first Arab businesses in adjacent streets.

The increased number of stores and the ethnic composition of dwellers were not regarded as significant social issues until the late 2000s (Venkov, 2012). Then, the perception and definition of this market modified significantly in terms of the revaluation of the urban space and the adoption of capitalist-based trade, consumption and urban development models. At the same time, citizens, administrative decision-makers and the media insisted on the need to refurbish and modernise the market through a 'Europeanisation' process; back then, Sofia's chief architect described the market as an 'unacceptable oriental landscape' (Stoyanova, 2008).

The general objective of this chapter, therefore, is to explore the connections that exist between the conflictive construction of identities and the gentrification processes promoted by the administrative sphere (which are supported by a group of local dwellers). This text focuses on the area used by the market and its surrounding areas to analyse the different opinions of relevant actors about ethnic identities and the multiculturalism of central Sofia. To this end, three different observation spaces were identified: the area that houses fruit and vegetable stores, which was refurbished in 2006, the new food kiosks built in 2014 and Tzar Simeon Street, where the vast majority of businesses are run by Arab or Central Asian nationals (most of them from Iraq, Syria, Afghanistan, Iran and Lebanon).

The specific objectives of this chapter are associated with the analysis of the different points of view about the cultural diversity of the market and its transformation in recent years. Special attention is given to the different opinions of relevant actors, which range from the nostalgic search for authenticity and Balkan culture to the selective adoption of a 'European' approach through the implementation of gentrification projects and the removal of certain ethnic groups from the central area of the city. In this sense this text understands sociability within the market context as a complex and constantly changing process, which may lead to the emergence of identity-related conflicts.

Different ethnography-based qualitative research techniques were used for the development of this chapter. This research was divided into different fieldwork stages – the last being carried out from May to June, 2016 – involving participant observation, interviews and collection of secondary data. Different interviews were conducted with customers, traders and workers of the market. The last of these stages was mainly focused on the workers and owners of the businesses located along Tzar Simeon Street, which is famous for its Arab stores. The inclusion of this street – which is not part of the market – is based on the negative opinions different interviewees have expressed about this space, particularly with a bias against certain ethnicities. Such a trend has increased over the last years in parallel with the increasing numbers of Arab immigrants and their visible presence in the area.

It is worth mentioning the sense of 'not-belonging' to the market felt by some individuals who declined to participate in these interviews. In contrast to the latter, and despite the indifference and hostile opinions about the market, participant observation enabled the identification of support and solidarity practices, or at least tolerance, towards these groups.

Understanding the markets of the 'East of the West'[2]

Eastern Europe has a wide variety of markets, which are defined according to geographic, economic and cultural conditions. Despite this, it is still possible to identify some common characteristics among these spaces. Different authors (Hüwelmeier, 2015; Polese and Prigarin, 2013; Sik and Wallace, 1999; Verdery, 1993) agree on referring to markets as key elements within the economic, political and social changes undergone by the Eastern Bloc since 1989. However, these markets have played different roles since their proliferation in the 1990s and they are now facing a stage of vulnerability and decline that may be connected with the neglect and renovation cycle common to commercial gentrification processes (González and Waley, 2013). On the other hand, it is worth noting the role played by the ethnic and Oriental characteristics attributed to Eastern markets. Before exploring this case study, this section analyses some of the specific features of Eastern European markets and certain characteristics they share with their Western counterparts.

During the second half of the twentieth century, the production and redistribution of food in the Eastern Bloc were controlled by the State, with the only exception being the small-scale production and sale of agricultural products (Kaneff, 2002). Food stores and the businesses located in markets were owned by the State and administered through public agencies and represented the official provisioning system. There were also some specialised and exclusive stores (Tasheva, 2016). However, Verdery (1996) highlights the importance of the presence of a 'second economy', which is described as a series of informal/illegal strategies where private individuals produced and/or sold goods and services by using the official means of production and resources provided by the State. This author also stresses the ethnic aspect of these economic exchanges, since regular customers not only belonged to the same circle of friends, but also to the same ethnic group. As for markets, this second economy operated according to a series of assumptions, some of them being the private production of agricultural products that were secretly sold to the closest circle of customers. The fall of the Eastern Bloc did not mean the end of this type of economy; on the contrary, this secret practice gave rise to the proliferation of markets and street markets during the 1990s and 2000s. Such a phenomenon was fuelled by trade liberalisation, a shortage of work and the combination of poor quality of life / lack of opportunities and the progressive adoption of a consumerist lifestyle (Sik and Wallace, 1999). These circumstances allowed the expansion of regular markets and the spontaneous and informal emergence of new markets; in both cases, they used the public spaces built during the socialist era such as squares, green areas or monuments (Petrova, 2011). This period was also characterised by a certain *laissez faire* approach adopted by the state, which was gradually withdrawing from production and redistribution activities. There was a high demand for affordable products and interest in generating income through the leasing or subleasing of commercial spaces.

The development and expansion of markets and private trade activities in Eastern Europe are regarded as a post-socialist phenomenon. However, our bibliographical research has shown that the study of markets in Eastern Europe has tended to associate them with the concept of the 'bazaar economy', coined by Geertz (1978) much more often than the concept of 'market'.

Geertz's work on the bazaar economy during the 1970s referred to a specific form of market organisation in countries such as Morocco and Indonesia. Hüwelmeier (2013, p. 52) in her work on post-socialist markets in Berlin, Warsaw and Prague chose this term to reinforce the idea proposed by Geertz, who defined bazaars as a 'distinctive system of social relationships centring around the production and consumption of goods and services'. In his work, Geertz identifies three major characteristics of bazaars: lack of information, customer retention and bargaining. However, these features are not exclusive to the global South. For instance, the audiovisual work conducted by Robles and Monreal (2008) in different markets located in Madrid, Valencia and Barcelona allows us to observe some of the characteristics identified by Geertz. Despite the formal, regulated and institutionalised nature of these markets, the daily activities recorded by the researchers show frequent trader–customer interactions that begin with the latter asking for the price or quality of products due to the lack of information. This audiovisual document also shows the flexibility of traders on the price of goods – an action reminiscent of bargaining – and the offering of discounts or gifts to retain customers.

However, within the context of Western Europe, this type of informality is not associated with the concept of bazaar, which is generally used in case studies about the reality of markets in post-socialist scenarios in cities such as Odessa (Polese and Prigarin, 2013) or Warsaw (Hüwelmeier, 2015). This practice defines the approach of social sciences to markets through the establishment of a division between East and West; such a method is based on the clear distinction described by Geertz: 'Bazaar, that Persian word of uncertain origin which has come to stand in English for the oriental market' (Geertz, 1978, p. 29).

The use of the concept of bazaar to refer to Eastern European markets might be interpreted as a form of orientalism which according to Said's definition refers to the process of 'dealing with the Orient by making statements about it, authorising views of it, describing it, by teaching it, settling it, ruling over it' (Said, 1979, p. 23). The association of Eastern European markets with the concept of bazaar and notions such as informality, uncertainty, clientelism and contrasting them with the more regulated markets in the West – supposedly governed by supply and demand and free competition models – can be seen itself as a form of orientalism. This is where the concept of Balkanism emerges, as a particular form of exoticism applied to the Balkans and its history, collective image and proximity to the Orient. According to Todorova (2009, p. 15): 'It is, thus, not an innate characteristic of the Balkans that bestows on it the air of mystery but the reflected light of the Orient. One is tempted to coin a new Latin phrase: "Lux Balcanica est umbra Orientis"' [The light of the Balkans is the shadow of the Orient].

Despite the differences in terms of the organisation, operation and understanding of Eastern markets, it is still possible to observe some similarities with their Western European counterparts. On the one hand, and in the same vein as previous cases in this book, it is possible to identify a certain 'devaluation' of markets as spaces for daily shopping and socialising. Within the context of a broader trend, which is associated with the transformation of consumption models and social networks, shopping centres are emerging as new spaces for leisure and social interaction (Sazonova, 2014). At the same time, new life has been breathed into markets through the organisation of events such as organic and eco-fairs and farmers' markets, which are intended to revitalise these spaces by disguising consumption as a unique leisure experience. Part of this trend is also associated with neo-traditionalism and the appeal to customer's emotions through self-identification with certain traditional products associated with the national identity.

The decline of markets may be related to the general arrival of supermarket chains and their marketing campaigns, which have been adapted and transformed over recent years. Initially, supermarkets promoted the consumption of Western products as desirable; today, however, major retail chains have adopted a more sensitive approach, and offer products tailored to each country or region (see for example an interview with Lidl Bulgaria's CEO, Capital.bg, 2016). The current strategy aims to generate nostalgia for traditional products or encourage experimentation and fusion between culinary cosmopolitanism and exotic ingredients.

As in the case of the other cities analysed in this book, this allows us to identify a double trend. On the one hand there is the re-emergence of events, fairs and 'conventional' or 'strip' commercial centres, which are targeted at the middle class and are built with the aim to recover the central area of cities for leisure activities; and, on the other hand, there are traditional markets, which are increasingly framed as obsolete, not only in terms of place, but also in terms of consumption, association and economic activity.

The redevelopment of the Women's Market in Sofia

This section offers a brief timeline of the transformations undergone by the central market of Sofia since the 1990s. These changes are not only related to infrastructure but also to the modification of symbolic spaces, especially in terms of space perception. In this case it is important to identify when and how the market began to emerge as a social issue, which went from being a needed and useful space to becoming the 'ghetto' or the 'ulcer' of central Sofia (this is how it is commonly referred to by the media, political officials and users of Internet forums).

During the 1990s, the Women's Market was defined by the characteristics associated with the boom and expansion of markets in post-socialist cities, which emerged as the consequence of the shortage of formal employment, the adoption of consumption as a lifestyle and the destigmatisation of trade as a private profit-making activity (Kaneff, 2002; Sik and Wallace, 1999). This space also served as a haven for excluded ethnic minorities, who were offered employment opportunities and inclusion. It is important not to idealise or attempt to identify harmonious and fluid social interactions; however, attention should be given to the fact that vendors, customers and poor individuals from different ethnic origins had also the right to remain and belong to a space in central Sofia through activities such as work, residence or education (in the case of the children of vendors).

In 2005, a group of neighbours founded a citizens' committee for the purpose of establishing a dialogue with the local administration on the reduction of the market's size. According to a letter sent to different institutions, this group expressed its concern about the expansion of the market and the negative consequences of such an action on the use of public space, the quality of life and the image of the neighbourhood. During its first years of existence, the 'Vuzrajdane' committee, as it was known, had no significant impact. Different measures were taken such as the collection of signatures and the reorganisation of the structure of the group; however, all requests made by the committee were unsuccessful. It was not until 2010 that a strategic alliance with mayoral candidates[3] enabled this group to achieve visibility and popularity. The committee held a series of meetings that were attended by local politicians and intellectuals; as a result, the opinions of this group about the condition of the market were heard and further disseminated.[4] Venkov (2012) analyses this conflict from the point of view of the success of a citizen participation campaign. However, the stigmatisation of the market would not have been possible unless the vendors were categorised as non-citizens.

This can be observed in the change of attitude of the media and users of Internet forums. Until 2004, the information available on the Internet was related to tips on how and where to find specific products and stores. However, since 2006, and especially since 2010, an increasing number of articles published by mainstream newspapers began to associate the market with crime, pollution and the inability of ethnic minorities to adapt themselves to what was regarded as acceptable and/or desirable for a European capital such as Sofia. During the last six years, the words 'ghetto' and 'orientality'[5] have been frequently used by the media to refer to the market. Some relevant examples are an interview given by the mayoress of Sofia to BTV, one of the main television channels of Bulgaria, entitled 'Women's Market, a ghetto in the center of Sofia' (Btv, 2011) or the article 'Life in the Women's Market, between a ghetto and a mall' (webcafe.bg, 2014), which discussed the effects of the renovation process in 2014, published on Webcafe, an emerging portal on current issues and leisure. On the other hand, the headlines pointed at the contradiction between the goal of Sofia to become a model of European city and its current reality: 'The markets: so much Orient within a supposedly European Sofia' (Sega.bg, 2010). Likewise, the Roma population has been frequently associated with concepts such as illegality, fraud and deception like in this article, entitled 'The Roma people control the "duty-free" tobacco business within the context of the street selling of non-tax-stamped cigarettes'.[6] This is a process similar to that described by Semi (2008) in the case of Turin, where illegal economic practices are intrinsically related to migrants because of their foreign status. The chief architect of the city also supports this type of discourse by referring to the oriental aspect of Sofia's centre as 'unacceptable'.

These circumstances led to the approval of the renovation project in 2012, within a context marked by a combination of negative discourses on space (promoted by *respectable* neighbours), politicians with electoral aspirations and the intervention of the city council's chief architect, who has decision-making power on urban transformation. The incorporation into the EU and the discourses on the 'return to Europe' had also a great impact on the perception of the identity of the city. Both the local administration and private investors aim to turn Sofia into a city model and a European centre through the implementation of an elitist and exclusionary project. The exchange value of land and real estate interests in the area surrounding the market are prevailing over the use value of land through the stigmatisation of poverty and its association with ethnic-related issues. An example of this is the list of requests submitted by the citizens' committee to the City Council, which demands the rights of proprietors and a duty for the administration to increase the value of real estate around the market area. Equally essential was the agreement on heritage demands, which aimed at recovering the old spirit of Sofia through the rehabilitation of dwellings located around the market. 'Rescuing Old Sofia' has been a recurrent topic on the press and a concern for different campaigns and pieces of research (Fading Sofia, n.d.).

The popularity gained by the discourses on the marginalisation of the Women's Market is also associated with the emergence of xenophobic parties (Ghodsee, 2008). This phenomenon is characterised by the convergence of concepts such as nationalism and 'Europeanism' which, according to Latcheva (2010), gave rise to a new sense of belonging and national pride through the exclusion of certain groups. In this way, wide sectors of the population – most of them well-represented in the market – are symbolically deprived of their national citizenship and European status. Therefore, the voices, lives and images of vendors and customers who want to preserve the market are practically ignored by the media and city decision-making processes. Similar to the discussion around the redevelopment of Queen's Market in London (Dines, 2007 and González and Dawson, this book), the renovation of the Women's Market does not include the human dimension.

The images that the planners developed regarding the renovated market show how architects have come to ignore the habitability conditions of the spaces they design and also represent the targeting of new users for the future market. There was little information available for traders and customers on the reconstruction of this space and there were no debates on this problem other than those held by the citizen committee; the local administration remained silent on this issue.

An exception was the debate organised by the architects of *Grupa Grad* (City Group), which took place in the market. With no rules other than respect, each participant expressed their opinions and doubts about the future of the market as may be seen in the video they recorded (Grupa Grad, 2015).

Figure 10.2 Part of the 'Women's Market' showing the new kiosks which replaced the
former market stalls removed in 2013
Photograph: Stoyanka Andreeva Eneva.

Despite failing to stop the reconstruction project and the lack of a genuine
dialogue with members of the local administration – who did not participate
in the debate – this event successfully managed to spread the personal and
social situations of affected people and the different interests involved in the
future of the market.

Finally, the reconstruction process was carried out during the 2013–2014
period and, unlike the former open-air stores, the new kiosks were used for different purposes (see Figure 10.2). The sale of fruit and vegetables was banned in
the new segment of the market; this measure forced former traders to move to
other areas of the market or sell new types of goods. On the other hand, the
increase in rents also caused the displacement of traders. Since the price per
square metre of the new kiosks was higher than that of former shops, some
traders were no longer able to stay in the market. These pieces of evidence
reveal the presence of a gentrification process: an increase in the price of
leasing fees, displacement of traders, physical renovation and the transformation of most commercial spaces into recreational areas. There was also a
symbolic transformation which took place through the creation of a discourse
on the 'new and European aspect' of the market. However, it is difficult to
find positive opinions about the renovation of this space from traders, customers
and the media – who firstly created a sense of urgency about the reconstruction of the market. The media criticised the delays, cost and aesthetic result of
the project; however, criticism was also levelled at the inability of the project to
'civilise' the area; There are still articles warning of the insecurity of the
neighbourhood and revealing that undesirable people have merely moved to
the Arab shopping area, which is referred to as 'Little Beirut' or 'Little
Baghdad' in order to highlight ethnic identities that do not fall within what is
'normally' regarded as Bulgarian. Finally, the media have expressed a growing

concern about the destruction of the market in terms of socialisation and daily shopping activities and the subsequent loss of employment for highly vulnerable people (Dnevik.bg, 2014). Soon after the reconstruction, recurrent headlines appeared, such as 'Neither for women, nor a market' (Capital.bg, 2014) and 'Could the modification of functions lead to the end of the Market?' (Bnr.bg, 2014).

The Women's Market and the role of ethnic diversity in its daily operation

Sociability is one of the characteristics that most attention has drawn within the context of the study of markets. However, when it comes to ethnic and social diversity, there are different opinions about the potential of markets to encourage social and cultural exchange. According to Hann (1992), the open and informal nature of these spaces enables the convergence of individuals from different backgrounds; However, such a mixture is not enough to evolve into cohabitation. Smith (1965) points out that the multiculturalism of markets may be referred to as 'mutual avoidance'; on the other hand, Watson (2009) suggests that these circumstances promote situations of 'rubbing along'.

In relation to these presumptions, the following points analyse the role of ethnic diversity not only in the functional operation of the Women's Market, but also in its capacity to create a multicultural space within the context of a gentrification process. This diversity however is expressed in very different and sometimes contradictory ways:

1   The market sometimes has the appearance of a relatively peaceful and tolerant space. This is related to the possibility of acquiring a wide array of affordable products independently of the nationality/ethnicity of the vendors. An example of this is Adam, who purchases products for his restaurant from Arab, Roma, Turkish and *ethnically neutral* Bulgarian traders. He buys from the market, serves to Arab traders and acts as neutral actor because of his religious observance, language proficiency and likeability. Due to his ability to get along with others and his interesting life story, this person is frequently interviewed about the issues and the future of the market and its surrounding area, thus becoming an informal spokesperson.

2   The self-exoticisation of Balkan culture expressed by the interaction with others which is manifested in the pre-redevelopment calls for preservation and defence of the market as well as in current feelings of nostalgia that the market conveys. This type of sense of belonging towards the market is not necessarily linked to the possibility of finding different and affordable products, but also with its lively atmosphere, which is described by Venkov (2012, p. 10) as a form of 'autochthonous, typical and Balkan exoticism'. This perception of the market is common among the post-1989 generation. These are people who have travelled, studied or worked in different parts of Europe or the US and associate this market with

regional and national authenticity and self-representation. Despite not being part of their lives or daily activities, the market reflects their Balkan, Slavic and non-Western identity and represents an opportunity for them to identify themselves with these ethnicities.

3   The emergence of neo-traditionalism: In this case the market is regarded as a picturesque place where it is possible to find traditional Bulgarian goods and artisans associated with activities such as pottery, crafts and food and cosmetic products. As of May 2016, the 'normatively Bulgarian' nature of this space can be easily identified in the reconstructed area. Some examples are the dairy store Bulgaricus Bg and the stores named after the surnames of their owners – Gavrilovi, Simeonovi – which is a measure intended to reinforce the local identity and presence in the area. The objective of these initiatives is to represent authenticity, which is also reinforced through the concentration and visibility of different craft and folklore stores in the central segment of the new market. This new approach – targeted at tourists through the display of graphic images of Bulgarian and traditional culture – is a key element of the renovation project. The creation of a pedestrian street and the opening of cafes and craft shops was the pinnacle of this project, prompting the media to make enthusiastic comparisons between this city and other European capitals. As for the recovery of old Sofia, there is KvARTal, an urban art festival that evokes the old spirit of the neighbourhood; however, despite being held in adjacent streets, this event's map ignores the presence of the market.

4   The emerging proliferation of minority ethnic businesses. In this context, importance is given to turkish delight shops, bakeries and Arab food shops, which are growing in popularity among local dwellers, regarded as 'pioneers' (Smith, 2012) as they dare to explore this part of the city. There is also the Multi Culti Map project, which aims at creating positive opinions about the ethnic and culinary diversity of the area. This initiative is based on the identification of restaurants run by foreigners, especially in the city of Sofia, and is not intended to create a culinary map of the area but to choose different establishments according to the project's team impressions and preferences. In general terms, this map is targeted at Internet-savvy and tolerant young users who enjoy cosmopolitan cuisine. Likewise, attention should be given to the comments of some of the media – mostly targeted at upper-middle-income users with cosmopolitan tastes – concerning the market and its surrounding areas. The latter is related to the search for exoticism and the use of headlines such as 'A scent of saffron and nostalgia' (Capital.bg, 2015) or descriptions that compare the market area with the streets of Havana 'with its old cars and cobblestones and a little reminiscence of Istanbul' (Webcafe.bg, 2015) in an article which portrays some of the men working on the Women's Market

The above classification shows the different opinions expressed by the media, neighbours and clients; however, the situation of the market according to traders and workers can be referred to in terms of coexistence rather than cohabitation among different ethnic groups (Giménez and Malgesini, 1997). This is mostly due to the lack of communication among these communities. There is practically no relationship between the market and its adjacent streets dominated by Arab stores. All Arab interviewees – either workers or traders – live in other neighbourhoods. If we consider the capital city as a whole, it is normal for some individuals to live away from their workplaces; however, it is important to pay attention to the stigmatisation of the zone as a residential space.

Stores are mostly perceived as workplaces and, to a lesser extent, as social spaces. On the other hand, streets are used for socialising purposes, as these are places where people meet to interact during breaks, when looking for employment or just to hang out. This type of sociability and street-based leisure ('urban bustle' as discussed in Chapter 11 in this book on markets in Quito) tends to be stigmatised within the context of the modern, European and efficient city model, where the non-commercial use of streets is constantly frowned upon (Hernández and Tutor, 2015). But, these relationships are limited and locally-based. Apart from Adam, who purchases products for his restaurant in fruit and vegetable stores and neighbouring butchers' shops, the rest of Arab men who took part in interviews have no relationship with the market. Their routines do not include activities such as doing the grocery shopping or cooking, as they go to Turkish or Arab restaurants, like Adam's restaurant. This situation and the distrust generated by people speaking in other languages hinders the building of relationships between this group and local dwellers. Therefore, they meet among their own countrymen and do not feel welcomed or encouraged to speak with local people and thus aspire to delegitimise the area and fight against gentrification.

On the other hand, high staff turnover (described by Adam and Carlos)[7] hinders the creation of a sense of identity and permanence or the collective association of traders. Having said that, the case of Barcelona demonstrates that illegal trade can generate collective organisation and politicisation in protest at the imposition of barriers, the Foreign Nationals Act, racism and labour injustice.[8] However, in this case study, and according to Venkov (2012), traders were not able to agree on common goals in order to fight against the reconstruction project. This has been due largely to the multiple tensions between discriminated and oppressed groups such as the Roma, Turkish and Arab people and old migrants or between economic and political actors.

Despite these particularities, it is possible to draw some parallels with the other markets described in this book in terms of the expulsion of diversity in cases where it is considered that there are no profitable or beneficial results. Gentrification processes do not usually start with the direct intention to attack and destroy markets, but to create new tastes, consumption habits and desire within the context of a city that has turned these spaces into undesirable objects. In Sofia, concepts such as disorder, noise and informality are now regarded as atypical, shameful and unworthy of an EU member country. In this way, the topophobic discourses justified and demanded the displacement of traders, poor people and individuals from other ethnicities. However, the reconstruction of the market represented an architectural quick-fix measure that failed to solve or even address the social problems of the area.

Conclusions

Eastern European markets have undergone a series of transformations over the last 25 years, from an initial proliferation and intense activity associated with informality to decay and neglect on the part of the local administration, including the current impetus for transformation and building European capitals and global, competitive cities that have no space for the informality and precariousness of surviving markets. This last phase has been defined by the re-emergence of nationalism and the development of a discourse and a sense of belonging to Europe that excludes those who do not fit with the national prototype.

This chapter has reviewed some trends related to the perception of ethnic diversity in markets by using the city of Sofia as a case study. The market is a space where different emotions such as rejection, fear, fascination and culinary exoticism converge. In this sense, the main contribution of this chapter is the analysis of the role of ethnic diversity within a context marked by the vulnerability of the market. Special attention is given to how the identity of people has been used to elaborate the image of a market that went from being useful, essential and characteristic of Sofia to an undesirable, avoidable and dispensable place. European aspirations, an increasing nationalism and the impetus of a group of neighbours to recreate 'the old Sofia' constructed a discourse centred on the urgency of transforming a central area of the city that has not been directly contested by market traders and customers.

However, even after the destruction of the market and the construction of modern buildings that could not be used by former traders as the result of their expensive leasing fees, diversity is still there, hidden among the borders of the new market area.

Notes

1 Original in Spanish translated by Juan Pablo Henríquez Prieto
2 This title was taken from the book *East of the West: A Country in Stories*, written by Bulgarian author Miroslav Penkov. In this work, the author thinks through and rethinks Bulgaria according to his migratory experience in the United States.
3 Meeting between local neighbors and municipal candidate Proshko Proshkov, who supported the destruction of the market. As for local vendors, the candidate said 'these people will leave after the completion of the reconstruction process. They will not feel comfortable within a clean environment'. Source: https://www.youtube.com/watch?v=grvN_67CqPM

4   Collection of signatures for the reconstruction of the market. Text put out by the citizen committee and published on the website of one of the political parties that supported such an initiative. Mayoral candidate Proshko Proshkov is a member of this party. Source: https://mydsb.wordpress.com/2010/04/29/dsb2010042901/
5   Interview with the chief architect, who referred to the old market as an 'ulcer' and promotes the renovation of such a space by stating that there is no need to discuss the implementation of this initiative since 'we have all seen the same market over the years.' Source: http://focus-news.net/opinion/2014/06/08/28712/arh-petar-dikov-rekonstruktsiyata-na-zhenski-pazar-shte-doprinese-za-po-dobriya-oblik-na-sofiya.html http://www.webcafe.bg/webcafe/reportazh/id_1869583556_Jenskiyat_pazar_-_mejdu_getoto_i_mola_
6   Roma people control the 'duty-free' tobacco business within the context of the street selling of non-tax-stamped cigarettes. Source: http://www.segabg.com/article.php?sid=20090817000040001401
7   A Peruvian trader who has been worked in the market for more than 20 years.
8   An excellent example of this is the Popular Union of Street Vendors of Barcelona.

References

Apostolova, R. (2014). From real socialism to real capitalism: The making and dismantling of the Vietnamese worker in Bulgaria. In: Apostolova, R., Deneva, N. and Hristova, T. (Eds.), *Situating Migration in Transition: Temporal, Structural, and Conceptual Transformations of Migrations. Sketches from Bulgaria*. Sofia: KOI Books, pp. 13–41.

Bnr.bg website (2014). Could the modification of functions lead to the end of the Market?14 July 2014. Available from: http://bnr.bg/horizont/post/100448766/izchezva-li-jenski-pazar-kato-promena-funkciata-si (accessed on 16 May 2016).

Btv.bg website (2011). The Women's Market, a ghetto in the center of Sofia. Interview given by the Mayoress of Sofia, 29 June 2011. Available from: http://www.btv.bg/video/1360840214-Jenskiyat_pazar__geto_v_tsentara_na_Sofia.html

Capital.bg website (2014). Neither Women's nor a Market, 11 July 2014. Available from: http://www.capital.bg/politika_i_ikonomika/obshtestvo/2014/07/11/2342042_nito_e_jenski_nito_e_pazar/

Capital.bg website (2015). A scent of saffron and nostalgia, 20 May 2015. Available from: http://www.capital.bg/politika_i_ikonomika/obshtestvo/2015/03/20/2496208_s_duh_na_shafran_i_nostalgiia/

Capital.bg website (2016). Interview with Lidl's CEO, 12 August 2016. Available from: http://www.capital.bg/biznes/media_i_reklama/2016/08/12/2810784_kak_bananite_za_69_st_stanaha_marokanski_podpravki

Dines, N. (2007). *The Experience of Diversity in an Era of Urban Regeneration: The Case of Queens Market, East London*. Milan: Fondazione Eni Enrico Mattei (FEEM).

Dnevik.bg website (2014). The Women's Market: New image, old issues, 3 September 2014. Available from: http://www.dnevnik.bg/gradska_sreda/2014/09/03/2373140_jenskiiat_pazar_-_nov_oblik_stari_problemi/

Fading Sofia Website (n.d.). Available from: http://www.fadingsofia.rcss.eu/

Geertz, C. (1978). The bazaar economy: Information and search in peasant marketing. *The American Economic Review*, 68(2), 28–32.

Giménez, C. and Malgesini, G. (1997). *Guía de conceptos sobre migraciones, racismo e interculturalidad*. Madrid: La Cueva del Oso.

Ghodsee, K. (2008). Left wing, right wing, everything: Xenophobia, neo-totalitarianism, and populist politics in Bulgaria. *Problems of Post-Communism*, 55(3), 26–39.

González, S. and Waley, P. (2013). Traditional retail markets: The new gentrification frontier? *Antipode*, 45(4), 965–983.

Grupa Grad (2015). Video of meeting of citizens discussing their right to quality of life in the market neighbourhood, 17 July 2015. Available from: https://www.youtube.com/watch?v=VsHoXZiem_U

Hann, C. (1992). Market principle, market-place and the transition in Eastern Europe. In: Dilley, R. (Ed.), *Contesting Markets*. Edinburgh: Edinburgh University Press, pp. 244–259.

Hernández, A. and Tutor, A. (2015). Espacio público: Entre la dominación y la(s) resistencia(s). In: Aricó, G., Mansilla, J.A. and Stanchieri, M.L. (Eds.), *Mierda de Ciudad. Una rearticulación crítica del urbanismo neoliberal desde las ciencias sociales*. Barcelona: Pol·len Edicions, pp. 58–72.

Hüwelmeier, G. (2015). From 'Jarmark Europa' to 'Commodity City.' New marketplaces, post-socialist migrations, and cultural diversity in Central and Eastern Europe. *Central and Eastern European Migration Review*, 4(1), 27.

HüwelmeierG. (2013). Postsocialist bazaars: Diversity, solidarity, and conflict in the marketplace. *Laboratorium Journal of Social Research*, 5(1), 52–72.

Kaneff, D. (2002). The shame and pride of market activity: morality, identity and trading in postsocialist rural Bulgaria. In: Humphrey, C. and Mandel, R. (Eds.), *The Market in Everyday Life: Ethnographies of Postsocialism*. London: Berg Publishers, pp. 33–51.

Kostentzeva, R. (2008). *Sofia, My Native Town*. Sofia: Riva editions.

Latcheva, R. (2010). Nationalism versus patriotism, or the floating border? National identification and ethnic exclusion in post-communist Bulgaria. *Journal of Comparative Research in Anthropology and Sociology*, 1(2), 187–215.

Petrova, V. (2011). Take the market out of sight. *Seminar BG*, 11(1), 2(1). Available from: http://www.seminar-bg.eu/spisanie-seminar-bg/special-issue1/item/319-take-the-market-outof-sight.html

Polese, A. and Prigarin, A. (2013). On the persistence of bazaars in the newly capitalist world: Reflections from Odessa. *Anthropology of East Europe Review*, 31(1), 110–136.

Robles, J. and Monreal, P. (2008). Mercados, vidas y barrios: Madrid, Valencia y Barcelona. Available from: https://www.youtube.com/watch?v=9qqUdB8O3co (accessed 6 September 2016).

Said, E.W. (1979). *Orientalism*. New York: Vintage.

Sazonova, L. (2014). Cultural aspects of sustainable development: Glimpses of the Ladies' Market. Friedrich-Ebert-Stiftung, Bulgaria Office, Analysis 2014. Available from: http://library.fes.de/pdf-files/bueros/sofia/10911.pdf

Sega.bg website (2010). The markets: So much Orient within a supposedly European Sofia. 26 April 2010. Available from: http://www.segabg.com/article.php?sid=2010042600040001401

Semi, G. (2008). 'The flow of words and the flow of value': Illegal behavior, social identity and marketplace experiences in Turin, Italy. In: Cook, D. (Ed.), *Lived Experiences of Public Consumption*. London: Palgrave Macmillan, pp. 137–157.

Sik, E. and Wallace, C. (1999). The development of open-air markets in East-Central Europe. *International Journal of Urban and Regional Research*, 23(4), 697–714.

Smith, M. G. (1965). *The Plural Society in the British West Indies*. Berkeley, CA: University of California Press.

Smith, N. (2012). *La nueva frontera urbana: ciudad revanchista y gentrificación.* Madrid: Traficantes de Sueños.

Stoyanova, S. (2008). Parks instead of markets in the City Centre. Press release, Sofia City Council, 27 November 2008. Available from: http://sofia.bg/pressecentre/print. asp?open=9&sub_open=36250

Tasheva, M. (2016). Retail gentrification and urban regeneration of the city of Sofia. Retrospective and perspective. Working paper series, Series (IV-4A) Contested Cities, Stream 4: Gentrification, WPCC-164018. Available from: http://contest ed-cities.net/working-papers/2016/retail-gentrification-and-urban-regeneration-of-the-city-of-sofia-retrospective-and-perspectives/

Thuen, T. (1999). The significance of borders in the East European transition. *International Journal of Urban and Regional Research*, 23(4), 697–714.

Todorova, M. (2009). *Imagining the Balkans.* Oxford: Oxford University Press.

Venkov, N. (2012). The civil society and the Women's Market. Research Centre for Social Sciences. Accessible online on http://rcss.eu (via login) or on the author's academia.edu site: https://www.academia.edu/1609895/%D0%93%D1%80%D0%B0%D0%B6%D0% B4%D0%B0%D0%BD%D1%81%D0%BA%D0%BE%D1%82%D0%BE_%D0%BE %D0%B1%D1%89%D0%B5%D1%81%D1%82%D0%B2%D0%BE_%D0%B8_% D0%96%D0%B5%D0%BD%D1%81%D0%BA%D0%B8%D1%8F_%D0%BF% D0%B0%D0%B7%D0%B0%D1%80_Civil_society_and_the_Womens_Market_

Verdery, K. (1993). Ethnic relations, economies of shortage, and the transition in Eastern Europe. In: Hann, C.M. (Ed.), *Socialism: Ideals, Ideologies, and Local Practice.* London: Routledge, pp. 172–186.

Verdery, K. (1996). *What Was Socialism, and What Comes Next?*Chichester, UK: Princeton University Press.

Watson, S. (2009). The magic of the marketplace: Sociality in a neglected public space. *Urban Studies*, 46(8), 1577–1591.

Webcafe.bg website (2014). Life in the Women's Market; between a ghetto and a mall, 15 July 2014. Available from: http://www.webcafe.bg/webcafe/reportazh/id_ 1869583556_Jenskiyat_pazar_-_mejdu_getoto_i_mola

Webcafe.bg website (2015). Men of the Women's Market, 14 Aug 2015. Available from: http://www.webcafe.bg/webcafe/reportazh/id_757807555_Majete_na_Jenskiya_pazar_

11 Popular culture and heritage in San Roque Market, Quito[1]

Eduardo Kingman Garcés and Erika Bedón

Introduction

This study aims to show the complexity of a 'popular'[2] economy and culture within different urban contexts (especially in commercial areas) and how this popular economy is increasingly entering into conflict with security policies and the police and well as with an exclusionary aesthetic associated with the notion of heritage. By popular economy we therefore refer to the relationships formed from exchange activities, that we call 'urban bustle'[3] or 'parallel forms of circulation characterised by constant urban-rural flows, the overlapping of formal and informal economies and a certain degree of autonomy in relation to capital and state action' (Kingman Garcés and Muratorio, 2014, p. 9). In order to capture this urban bustle we have developed a long-term investigative strategy and in this chapter we present our research in the market and neighbourhood of San Roque in Quito. The empirical references used in this chapter are based on an ethnography to study the social memory of this part of the city through interviews, formal and informal conversations and field observations conducted to understand the social architecture of this neighbourhood and its long-term composition. It is worth noting that the reflections on the dynamics of trade in Quito – referred to in this study as 'urban bustle' – are based on long-running historiographical work in interaction with the current urban renewal and population displacement processes. Part of our methodology consists in relating events that take place in different periods of time according to a single paradigmatic perspective. Likewise, we also seek to establish a relationship between apparently independent phenomena such as heritage and the police with the memory of these social sectors and different urban planning policies.

This chapter explores three analytical areas that have already been discussed in the Introduction to this book, which interact with other chapters: 1. Processes of physical obsolescence and territorial, class and racial stigmatisation suffered by the Mercado de San Roque, a place which has been historically neglected by public authorities. More recently this area has been targeted by state intervention in the form of urban renewal strategies based on notions such as security, administration and population control. The study of such security measures and police actions allow us to identify the different socialisation and resistance practices of groups using the market and their relationship to the city. 2. The disputes among public interventions, heritage-led urban renewal and real estate speculation in relation to the San Roque neighbourhood and the Mercado de San Roque. 3. Additionally, this market is further analysed according to the views of indigenous and mestizo people, who associate this area with hospitality and daily life.

Mercado de San Roque and the San Roque neighbourhood: Between abandonment and renovation

San Roque is a neighbourhood located in the historic centre of Quito, the capital city of Ecuador. This area, which is yet to be fully integrated in the formal urban fabric of the city, is characterised by its market and the vast majority of its population is made up of indigenous and low income people.

The Mercado de San Roque offers both wholesale and retail products and it is also a space where the local social fabric of the neighbourhood is woven. According to the media, both the market and the neighbourhood are run-down, dirty and dangerous areas that require intervention. With time, such an opinion has entered into the 'common sense' of citizens of the wider Quito. In line with this thought, San Roque is both polluted and polluting, concentrating some of the most stigmatised areas of the city: the market, the local prison and sex work areas and a segment of the population that is regarded as a marginalised or 'pariah' community.

Such stigmatisation has triggered concerns from local public authorities about the support and development of these areas. This is a growing issue shared by other Latin American cities (Caldeira, 2007; Delgadillo, 2016). What is noteworthy in these cases, is how this renewed attention is however placed within a longstanding neglect for low income settlements in the cities. In the case of San Roque, little attention has been given to the area's environmental, economic, social and security conditions. However, this area has suddenly drawn the attention of the local municipality, the state and its citizens. These largely ignored, abandoned and neglected spaces are now a matter of interest and concern. Our understanding is that the recent interest in these deprived neighbourhoods is a cynical concern associated with urban and security policies. In this sense, we seek to understand the factors that turn a specific place, such as a public market, into a vulnerable, violent and denuded space and why this space – unlike other ignored areas – is favoured by intervention and development measures. Likewise, attention is given to identifying the meaning of this new kind of concern.

This analysis suggests that the interest in sites like San Roque is also related to the proximity of these places to the areas of the city recently renewed, particularly those with a heritage value. Reflections on this issue not only shed light on the perceptions of citizens about popular neighbourhoods, but also help us to understand the organisation of the city: on the one hand, large popular zones are ignored as the result of divisions within the city. On the other hand, there is a specific concern about certain areas in terms of renovation, gentrification and heritage.

Some spaces of Quito and other Andean cities have historically been associated with the emergence of borders between the city, the countryside and the urban-rural periphery. We understand the concept of border to mean spaces of encounter, relationship as well as conflict (Kingman Garcés, 1992). San Roque has been regarded historically as the point of arrival and relationship with the city and also as the point where urban and rural spheres converged. In this sense we cannot speak about these border spaces in the *modern* sense of public spaces but as *frontier-spaces*. This frontier dimension was not only associated with trade but also with *shared customs*. Neighbourhoods in proximity to each other such as Santo Domingo, San Francisco, San Sebastián and San Roque played a key role in the development of a 'popular' economy, religion and culture that was in between the rural and the urban. By popular culture we understand a particular way to feel and perceive things that have emerged from the mixture of and relationship between different social classes and sectors over time. For example, Muratorio (2014) shows the role of the *Cajoneras* in Quito, women who sold hand-made produce (haberdashery, dolls, accessories) from chests of drawers in local squares, in the development of forms of popular consumption and in the reproduction of particular forms of aesthetics and taste.

The location of what has until recently been most important market of the city (near San Roque Church) gave a particular feel to the neighbourhood. During the 1950s and 1960s, this area became a large space that divided as well as connected the rural and indigenous world to the urban dynamic. The neighbourhood was home to the market, grocery businesses and stores focused on the selling of religious images and renting of costumes for the feasts of Corpus Christi and the Holy Innocents. It also hosted the former San Juan de Dios Hospital and the municipal prison and it was the final destination of interprovincial and inter-municipal transport routes, where it was possible to find accommodation, cheap restaurants, *chicherías*,[4] taverns, brothels, places for the sale and purchase of craft products, second-hand tools and used clothes. This space was also used for the recruitment of construction workers, carpenters, plumbers and workmen.

This neighbourhood gradually became a trading area and the residence of lower income groups in the context of a long-term transformation of the historic centre of Quito during the first decades of the twentieth century, which also involved a gradual change in the use of land, buildings, public spaces and the progressive migration of the elite. By this we refer to a long and ongoing appropriation process led by low income groups in neighbourhoods such as San Roque which, from 1950 to 1980, was turned into a shared, though culturally- and ethnically-differentiated, space. Trade, informal jobs and the occupation of the old houses once inhabited by the elite played a key role in this process. Today, most of the local population is of indigenous or mestizo origin. However, a significant part of the residents come from other low income urban neighbourhoods and even from middle income backgrounds; this circumstance turns San Roque into a border area.

During the last decade, this appropriation process by low income and indigenous groups has persisted mainly around the market area of San Roque. This process runs counter and is in conflict with a general trend of 'recovery of the historic centre' by the middle and upper classes and by public policies on heritage and tourism implemented in neighbourhoods such as La Ronda – as analysed by Lucia Duran (2015) – and San Sebastián also in Quito. This process is also taking place in other Latin American cities (Janoschka and Sequera, 2016) and affecting markets as discussed by Delgadillo in Chapter 2 of this book in the case of La Merced market in Mexico City, a stigmatised neighbourhood with an overwhelming popular trading culture that is threatened by renovation and tourist promotion in the historic centre of the city.

San Roque: A neighbourhood and market regarded as liminal, dangerous and stigmatised space

The Mercado de San Roque has been historically identified as a place where different populations converge and as a space that connects the urban and rural spheres (Minchom, 2007). The dynamics of popular life in Quito during the nineteenth and early twentieth centuries were marked by popular commerce, which allowed a certain degree of liberality in terms of the relationships among different social groups. Chances for social hybridisation were mainly based on these trade dynamics. Although it did not erase ethnic differences, the market enabled the emergence of relatively open spaces for daily exchange activities – the so-called 'urban bustle'. These occasional encounters were moments in which people talked to each other even once the exchange process was over. We could even say, extrapolating from Hardt and Negri's (2006) ideas, that this activity generated common elements through performative activities. One of the clearest examples of this performativity was associated with religious events and festivals, but also street trade and theatre, as well as moments of conflict and protest.

It is precisely this world of squares and streets open to multiple actors which began to crumble by the last third of the nineteenth and early twentieth century with the advent of modernity. The turn of the century witnessed the regulation of the market according to hygiene and *beautification* standards; however, these measures were not a technical order as much as a civilising initiative. *Beautification* meant the consolidation of the elites, contrary to the (bad) taste for baroque ornamentation common to popular traditional culture: motley altars, processions led by musicians and dancers and the clothes worn by the religious figures. Both beautification and taste were oriented towards social distinction, division and the construction of a model for progress. Within the context of a stratified society, modernity was particularly expressed through public representation. Events such as ceremonies and the honouring of awards, titles and ornamentation contributed to the reproduction of a hierarchical model in a secular context. However, this modernisation process was slow and did not reach the whole city but only operated in certain spaces.

This progressive spatial division and segregation has led to the current perception of low income neighbourhoods in Quito, which are regarded as dangerous spaces that should not be visited. The first indicators of this process emerged during the early twentieth century, at the dawn of modernity; however, such a phenomenon gained prominence towards the end of the century, when the southern area was stigmatised by security policies. It could be argued that these areas have been abandoned by the state, being neglected in terms of satisfying their basic needs and subject to 'low-level police' (in Rancière's (2006) terms) and arbitrary small mafias (Agamben, 2004). These spaces also represent the phenomenon of urban fragmentation common to postcolonial and late capitalist dynamics in which we see at the same time the rise of high-consumption areas fortified as safe zones (Caldeira, 2007) as well as areas with high levels of poverty. These fragmentations are a territorial expression of processes of economic deregulation and the rise of a supernumerary or superfluous population. The creation of these neighbourhoods of extreme marginality may be a factor in additional stigmatisation of groups of people, identified by Wacquant (2001) as being construed as 'urban pariahs'. The labelling of these neighbourhoods as dangerous places reinforces, in turn, the criminalisation of poor populations. This is a hegemonic imaginary partly created by the media, which is exacerbated in particular in those areas bordering regenerated or heritage zones. Our hypothesis is that the stigmatisation of specific places as dirty, dark and dangerous areas precedes the implementation of concrete security policies such as the deployment of petty police forces, 'neo-hygiene' and social cleansing.

The media constantly refers to high rates of insecurity and identified it as one of the most dangerous areas in the city. In 2003, the press reported that:

> According to an anonymous dweller, San Roque experiences a steep increase in crime each Tuesday, Friday and Saturday, which is when the local street market takes place. [...] The Municipality installed a surveillance camera on the corner of San Roque Church. However, there is not much crime in that area, so we asked for increased security in the market area and the tunnels used by criminals.
>
> (Diario Hoy, June 5, 2003)

The attention given to San Roque is a relatively new phenomenon that goes hand-in-hand with the intervention proposals made by recent municipal administrations. The exacerbation of fear is part of the initiative aimed at intervening in San Roque; this is conceived by the authorities and the media as a complex process given the strong social fabric built around the market but also an urgent process given the location of the neighbourhood, which is adjacent to areas that have received funds for heritage tourism purposes. For example, in the local newspaper the area was described in these terms:

> The historic centre of Quito is 300 hectares, including the colonial area and neighbouring districts. Despite successfully overcoming issues such as the occupation of streets on the part of vendors, the area is still affected by the presence of small criminal bands or criminal networks such as 'Mama Lucha', as well as problems of extortion, drug consumption and smuggling.
>
> *(Diario Hoy, June 5, 2003).*

It is important to highlight that the criteria for the intervention by the local municipality in San Roque are often led by security issues and not the need for basic services. For the National Police, crime in Quito is mostly based on the market area of San Roque. Therefore, efforts are focused on rooting out this illegal activity. Urban planning interventions are therefore conceived in terms of securitisation and social cleansing according to roughly four criteria: redevelopment and control of public space, the eradication of street vendors and 'urban bustle', the creation of organised and regulated markets without small and informal traders and even the gradual substitution of markets themselves by supermarkets, including in this retail model the so-called 'popular supermarkets'.

Paradoxically, the analysis of indicators of danger and georeferenced maps reveals that North of Quito is the most dangerous area in the city in terms of property crime and sexual violence. This area is also regarded as the most secure place in the city, even when the number of offences is three times higher than in the central area (Metropolitan Observatory for Citizen Security, 2013).

It is true that the historic centre is not free from being affected by violence and San Roque has to deal with everyday violence, however, these events occur much more frequently in other parts of the city. As Wacquant (2000, 2001) suggests, stigmatisation campaigns in certain areas also involve the criminalisation of poor populations and the instrumentalisation of security measures. As part of the security imaginary, the historic centre is seen as a place where stolen objects are traded, with San Roque being regarded as a crime hotspot. According to the institutional discourse, these illegal activities are supported by local vendors, who are associated with criminal networks or are relatives of those who trade stolen objects. Despite the rise of individualisation and the loss of social contacts common to late capitalism, the resilient social fabric is the focus of stigmatisation and criminalisation.

Urban renewal policies in San Roque

This stigmatisation process has preceded a series of urban interventions intended for cleansing and hygiene purposes. The implementation of urban renewal policies has been based on advances into areas so far not regenerated akin to colonial methods of conquest and establishment of liberated or recovered spaces. In the case of the city of Quito, this is particularly clear in the neighbourhoods of Santo Domingo, San Francisco, la Ronda, la Veinticuatro de Mayo and San Roque.

Our analysis of the historic centre and its limits reveals the presence of invisible thresholds and frontiers between heritage-led renewed areas and spaces yet to be intervened on. The role of the municipality is to organise these areas and implement regulation on land use. Though these initiatives appear to be urban interventions, they hide a social engineering purpose; that is, interventions on the local population.

In the case of Mercado de San Roque, local public authorities have proposed different intervention projects. One of the first intervention proposals was included within a comprehensive intervention plan for the historic centre of Quito in 2010. But due to the succession of local governments and the resistance of local vendors and dwellers, this project has not yet been fully implemented, suffering a series of modifications. However, there have been some partial reforms, especially when it comes to the negotiations with traders about their future relocation and the reorganisation of trade inside the market; these measures have been regarded by the authorities as a notable step forward within the context of this process.

In the particular case of San Roque, different municipal governments have tried to eradicate the market (or its conversion into a smaller neighbourhood market) and the whole social environment. Negotiations have therefore not been focused on the elimination of this market – this has taken for granted – but on the conditions underlying such a process. For the last few years, the municipality has proposed different relocation possibilities; however, these proposals have been rejected by the local population. One of these ideas consists in creating two markets – both in North and South Quito – to unite the wholesale markets based on San Roque, La Ofelia and Mercado Mayorista del Sur. This project also includes a series of traffic and mobility regulations so as to prevent heavy vehicles from entering Quito. Such a measure would interrupt the flow of products into San Roque, thus altering the commercial dynamics of the area – which is today regarded as a wholesale market – and controlling informal trade. According to a local municipal official:

> turning to the subject of the negotiation process, the Municipality seeks to relocate this market; first of all because it is located in the historic centre and there is the heritage discourse, there is no space for marginality in the centre of the city. Negotiations started in 2006, this relocation is part of a much larger process based on the relocation of the trading system of perishable goods; the problem is the presence of different markets and everything is a chaos, there is no control. Mercado de San Roque is planned to be turned into a neighbourhood market, but they do not want that because this is a profitable business for them, they do not need to

invest any sum of money and use the public space for free. This space will be recovered for the implementation of the urban project and the creation of an international Craft Centre; relevant actors are expected to become true artisan experts, the pieces of furniture made here [at the moment] cannot be regarded as craft items, they should have a series of unique characteristics.

(Interview with a local municipal official, August 2008)

Previous experiences of displacing informal trade from the historic centre over the last two decades suggest that the negotiation processes failed at an unexpected time, when the municipality took quick and irreversible actions. All of these *recovery* measures, as the authorities refer to them, have been combined with negotiations with traders and dwellers and then followed by unilateral interventions; this reveals that decisions were already made by authorities before consultations.

The case of Mercado de San Roque has been preceded by stigmatisation campaigns and discourses on public assets that have gradually permeated the common sense of citizens – especially those in charge of intervention measures. These urban actions also involve the implementation of State-led policies on social cleansing and the displacement of specific social sectors from urban renovation areas.

San Roque: A hospitable space

San Roque has been regarded as a dangerous place; however, our research suggests this area offers rich street-based relational spaces and urban hospitality (Azogue, 2012). Following Levinas (2001), we can understand this as a hospitable space, a meeting point for indigenous migrants within the context of a city where the presence of 'other' individuals is hardly accepted. The description provided by one of our interviewees could not be more eloquent: 'This is an indigenous space, a space that is home to indigenous migrant populations ... This is a space you are familiar with despite all the opinions against this area; in the end, this is a meeting space' (Interview with JC, Heifer-Flacso research group, July 2008). Here we are discussing a hospitable space that allows the development of particular forms of relationship in which indigenous people are main actors; such a situation is strengthened by the presence of the market as illustrated by one of our interviewees:

I think this occurs because of the presence of the market, this is a populated area, this is like a community space where we see each other every day, if you go to San Roque you will always be able to see an indigenous individual, someone walking around, people doing business over there ... the indigenous settlement located in the area may explain this situation.

(Interview with JC, Heifer-Flacso research group July 2008)

This space enables the emergence of different forms of aid and reciprocity that go beyond the individual sphere. In other words, these are collective forms of assistance and care, where the 'obligation to welcome newcomers is a moral standard' (Azogue, 2012, p. 23). These relationships also enable the generation of family and cooperative networks, which are essential for the survival and thriving of these families at any point in their migratory processes. The latter has a critical influence on the consolidation and maintenance of these networks. The senses and forms of appropriation that emerge in these spaces are diverse in nature and depend largely on the characteristics of human groups.

It is worth mentioning the appropriations and meanings given to these spaces by the indigenous women who are temporarily or permanently living in the city. Historically, there has been a strong presence of indigenous women who moved to the city to sell different products in streets and squares. Clorinda Cuminao (2012), in her research on Mercado de San Roque (and Manuela Camus (2002) – who focuses on the Guatemalan case) discusses how the spaces within the market – which are mainly regarded as female spaces – became consolidated as the result of a gender-based division of labour. For indigenous women, this space is a place where they can maintain and reproduce certain identity elements such as the way of dressing, their language and the care of children in the workplace. This has enabled them to develop a sense of pride and belonging, which is reflected and reinforced in the everyday spaces associated with religious and cultural activities that transcend the physical location of the market. Examples of these spaces are certain meeting points such as ballrooms or cheap restaurants, which are currently disappearing as the result of the implementation of heritage policies.

These life experiences allow us to regard San Roque as a space that favours the emergence of relationships that do not frequently occur in the rest of the city. Therefore, this place should not be referred to in terms of anonymity – though many, especially young people, prefer to remain unseen – but in terms of a relational space where it is common to see face-to-face contact among equals or those who pretend to be in the same situation of their peers, even when economic and social differences and power relationships are evident. Likewise, San Roque is not an unidentifiable or blank place, but an important and meaningful space. The latter is valid for indigenous newcomers and people from low income backgrounds who – despite not having indigenous roots – live in different parts of the city and identify themselves with San Roque. We are, then, observing a rich and clearly characterised space defined by the flow and circulation of predominantly indigenous people. This does not occur in other parts of the city, where relationships have become more extensive, diffuse and impersonal. This was expressed by one of our interviewees

> We lived in this sector along San Roque, la Magdalena and Cima de la Libertad [...] my sisters moved not too far away, to la Magdalena and la Mena but since their children are enrolled in a nursery school in San

Roque, the whole family moved there [...] since they have to do business and sell things, they go to San Roque, which were all indigenous people are. It could be said that San Roque is a space where families, divided into groups and distributed over different places, converge every morning. For instance, we used to have a space where we unloaded goods every morning and the whole family and the migrant community gathered there.

(Interview with JC. Heifer-Flacso research group, July 2008).

The market is not only the workplace of the indigenous population based in San Roque but also the space where these individuals establish their relationships with the city. They set up their businesses with effort, 'in an honest manner' (according to some testimonies), thus earning the respect of their counterparts. If the city excludes these new dwellers, they will redefine the sense of community within the urban space. This occurs in an everyday basis in the organisation of market activities, the construction of collective dwelling and especially when it comes to resisting and fighting eviction attempts. For example,

[V]endors are fighting to stay here, if they fought through demonstrations, spending nights on the streets in order to protect their businesses because if they were not there, their workplaces would be taken away from them, it would be difficult to evict them. Did you see how they defended their businesses? The municipal police came here and prevented people from going to their stores ... .

(Interview fish stallholder at the market, Mr. Alfonso R. José A. Heifer-Flacso research group. July 2008)

Though it is true that vendor associations have played a pivotal role in resisting regeneration and evictions, it is out of the question for the municipality that the area will be 'recovered' i.e. the market will be eradicated; such an objective will be achieved even by fragmenting local associations and establishing alliances with certain groups within the market in return for stores in the new market, or by convincing them that enjoying some benefits is better than losing all.

Religious celebrations also provide important spaces or moments for the generation of a sense of community inside the market area; these are significant events in the reproduction of culture as a *common place*, something that transcends the city and the symbolic occupation of the urban space. The organisation of these celebrations involves traditional knowledge and forms of representation; according to Mr. Andrango, trader, 'you have to know how to celebrate this feast' (R. Andrango, personal communication, May 2012). This implies the participation of local dwellers in the preparation of the event, the appointment of leaders, designate who is going to be in charge of preparing food, the elaboration of decorations, souvenirs, etc. The leaders of the event are responsible for elaborating the costume that will be used by the saint during the annual celebration. For instance, in Mercado de Iñaquito, women created a special space to keep the costumes and clothes used during the 'Divine Jesus' and 'Jesus of the Great Power' (patron saints of the market) celebrations; it is worth mentioning that such a room dates back to the foundation of the market. These costumes vividly preserve a popular aesthetic; likewise, this space is also used to keep the altar and other significant objects used during the event. These are occasions in which local dwellers use urban spaces as stages for the enactment of religious situations, where spiritual representations show us that the market is alive and full of senses and meanings. This space reproduces a series of *common customs* that enable the creation and exchange of economy and religious activities, thus revealing the urban–rural nature of popular culture.

The market parade, a traditional procession of the various markets to start off the annual Quito festivals, is another opportunity for the reproduction of social practices. The organisational level of this event transcends every market in the city and involves the presence of a series of family, commercial and work networks organised under the Market Federation or the Market Union. These organisational forms are testimony not only to the agency and degree of independence but also to their political nature. The different markets of the city should not be regarded as individual or unorganised spaces; on the contrary, their interconnection and forms of organisation are examples of their complex network of relations.

Conclusions

The purpose of this chapter was to show the relationships that exist between Mercado de San Roque and its neighbouring areas (San Roque neighbourhood) and the disputes over security, urban renewal and heritage policies in relation to this space. San Roque has historically been regarded as a relational space where job activities and popular commerce converge within the context of the so-called 'urban bustle'. While formerly composed of a mostly middle-class population, this neighbourhood is today home to indigenous and mestizo populations whose economic and social activities are directly associated with the market.

San Roque could be defined as an interstitial zone located between the urban and rural spheres, a space that enables the emergence of multiple socio-ethnic relationships and expressions of popular culture associated with urban requirements and communities. When seen as a frontier space, this area acts both as a meeting-dividing and an exclusionary-inclusionary point.

This market is characterised by the reproduction of different forms of everyday violence perpetrated by the police, criminals or even by exchange activities. These dynamics have been stigmatised as an excuse to justify the implementation of intervention policies. At the same time, for the 'popular' sectors that depend on the market for their daily subsistence – and especially for the indigenous individuals who live and work in San Roque – this area is regarded as an interstitial space where everyone 'speaks the same language' and where different housing, education and security practices are being implemented from a bottom-up approach. For many people from rural and indigenous backgrounds, San Roque is a hospitable place that offers new life opportunities and self-protection within the context of an inhospitable 'urban order', even if this situation involves subordination.

We have seen that heritage policies are not aimed at improving the habitat of low income groups located in the so-called historic areas; on the contrary, they neglect, degrade and stigmatise these spaces, justifying intervention on security, planning and urban renewal grounds. This should be associated with concepts such as population governance and management (Foucault, 2009).

This research has sought to reveal the hidden relationships that exist between heritage, the police and real estate investment. The stigmatisation of the market – orchestrated by the media – goes hand in hand with the conversion of this place into an object of desire by the heritage-related economic interests. However, despite multiple intervention attempts, municipal negotiations have met with failure.

Notes

1 Original in Spanish translated by Juan Pablo Henríquez Prieto
2 Editorial note: In original 'popular' in Spanish is a difficult concept to translate to English. Here it means an economy and a culture associated with low income groups. It can also refer to informal practices, those belonging to the 'people' in a way that differentiates them to groups in position of power, with economic wealth and part of the 'establishment'
3 Editorial note: in Spanish in the original *trajines*.
4 Places where *Chicha* (maize liquor) is sold. These used to be meeting spaces for indigenous people.

References

Azogue, A. (2012). El Barrio de San Roque ... Lugar de Acogida. In: Kingman, E. (Ed.), *San Roque: indígenas urbanos, seguridad y patrimonio*. Quito: FLACSO, sede Ecuador; HEIFER, Ecuador, pp. 21–35.
Agamben, G. (2004). *Estado de Excepción, Homo Sacer II*. Barcelona: Pre-textos.
Caldeira, T. (2007). *Ciudad de Muros*. Buenos Aires: Gedisa.
Camus, M. (2002). *Ser indígena en Ciudad de Guatemala*. Guatemala: FLACSO – Sede Guatemala.
Cuminao, C. (2012). Construcción de identidades de las vendedoras Kichwas y mestizas y los juegos de poder en el merca4o de San Roque. In: Kingman, E. (Ed.), *San Roque: indígenas urbanos, seguridad y patrimonio*. Quito: FLACSO, sede Ecuador; HEIFER, Ecuador, pp. 79–100.
Delgadillo, V. (2016). *Patrimonio urbano de la Ciudad de México. La herencia disputada*. México City: Universidad Autónoma de la Ciudad de México.
Diario Hoy (2003). Las redes de extorsión apuntan al centro de Quito. *Diario Hoy*, 5 July 2003.

Duran, L. (2015). Barrios, patrimonio y espectáculo: Disputas por el pasado y el lugar en el Centro Histórico de Quito. *Cuaderno urbano*, 18(18), 141–168. Available from http://www.scielo.org.ar/scielo.php?script=sci_arttext&pid=S1853-365520150001000 07&lng=es&tlng=es

Foucault, M. (2009). *Seguridad, territorio y población: Curso en el Collège de France: 1977–1978*. Buenos Aires: Fondo de Cultura Económica.

Hardt, M. and Negri, A. (2006). *Multitud*. Barcelona: Random House, Mondadori.

Janoschka, M. and Sequera, J. (2016). Gentrification in Latin America: Addressing the politics and geographies of displacement. *Urban Geography*, 37(8), 1175–1194.

Kingman Garcés, G.E. and Muratorio, B. (2014). *Los trajines callejeros: memoria y vida cotidiana. Quito, siglos XIX-XX*. Quito: Instituto Metropolitano de Patrimonio, Fundación Museos de la Ciudad.

Kingman Garcés, E. (1992). Ciudades de los Andes: homogenización y diversidad. In: Kingman Garcés, E. (Ed.), *Ciudades de los Andes. Visión histórica y contemporánea*. Quito: IFEA-Ciudad, pp. 9–52.

Levinas, E. (2001). *Entre nosotros: ensayos para pensar en otro*. Valencia: Pre-Textos.

Metropolitan Observatory for Citizen Security (2013). 20 OMSC Informe Estadistico y Georeferenciacion Octubre 2013. Quito: Author. Available from http://omsc.quito. gob.ec/index.php/biblioteca-virtual/informes-mensuales.html#

Minchom, M. (2007). *El pueblo de Quito. 1690–1810. Demografía, dinámica sociorracial y protesta popular*. Quito: FONSAL.

Muratorio, B. (2014). Vidas de la Calle Memorias alternativas: las cajoneras de los portales. In: Kingman Garcés, G.E. and Muratorio, B. (Eds), *Los trajines callejeros: memoria y vida cotidiana. Quito, siglos XIX-XX*. Quito: Instituto Metropolitano de Patrimonio, Fundación Museos de la Ciudad, pp. 113–148.

Rancière, J. (2006). *Política, policía, democracia*. Santiago de Chile: LOM Ediciones.

Wacquant, L. (2000). *Cárceles de la Miseria*. Buenos Aires: Manantial.

Wacquant, L. (2001). *Parias Urbanos: marginalidad en la ciudad a comienzos del milenio*. Buenos Aires: Manantial.

12 Conclusions

International perspectives on the transformation of markets

Sara González

Introduction

In this book we have critically analysed the state of traditional markets in Europe and Latin America and our cases show that traditional forms of retail are undergoing profound forms of transformation. We have seen that markets occupy a contested position in many cities as they bring together diverse and often contradictory interests and functions. They are often neither public nor private spaces; they can be strongly regulated by state and private authorities but at the same time informal practices arise and are maintained in the interstices between the formality. Markets have historically attracted vulnerable, low income and marginalised groups as places for work and to access affordable food and services and they are recognised as contributing to social inclusion. But this very same condition makes them vulnerable to gentrification processes where the informal atmosphere that sustains practices of solidarity can be eroded as traders and market users are displaced and replaced by much more corporate and regulated forms of retail.

In our book we have seen how these trends are apparent in many of our case studies regardless of their variegated geographical location, size, form or type of market. In the introduction of this book we signalled the difficulty of defining markets given their diversity and the specific qualities depending on their history and locations. Thus, markets are extremely varied and embedded in their location but at the same time there is something universal about them. Many of the traders, market managers and users that we interviewed in our research for this book said very similar things despite being located in very different cities and parts of the world. This suggests that the particular function they play in cities, described above, creates a similar social ecology. There are also macro-economic trends such as globalisation and the restructuring of the retail sector with the growth of internet shopping and international chain-shops and supermarkets that affect local retail. In addition, further tendencies in urban development such as gentrification, inter-urban competition, commodification and privatisation of public space increasingly affect most cities in the world. Thus, we find similar trends affecting the very diverse markets addressed in this book. In a similar way, Zukin et al. in their study of shopping streets in six cities around the world, find that 'neighbour-hood shopping streets are both extremely global and intensely local' (Zukin et al., 2016, p. 201).

In this concluding chapter we want to explore these trends and compare and relate the case studies that we presented individually in the book. In the introduction, we presented three analytical frameworks to study how and why markets become contested spaces: 1. markets as frontier for processes of gentrification; 2. markets as spaces for resistance and political mobilisation and 3. markets as spaces for the development of alternative practices of pro-duction and consumption. At that point, this analysis came from bringing together various existing literatures that touch on markets. In this conclusion we return to this framework this time drawing from our own case studies. As we already advanced in the introduction, these analytical frameworks are flexible and act as entry points into the various processes at work in each of the analysed markets, rather than as exhaustive explanatory models. I will now return to these analytical entry points to understand in more depth how the various processes affecting markets interact, blur, converge with or con-tradict each other. To do that, in the new few pages I focus on several pro-cesses and concepts that capture the main transformations occurring in our case studies markets.

Variegated gentrification

Processes of gentrification are definitely at play in many of the markets that we have discussed in this book. This is not necessarily surprising since gen-trification has been discussed now for a while as a 'global' process with a colonising force (Atkinson and Bridge, 2004) moving into cities all over the world. More recently Lees et al. (2016) have discussed gentrification as a planetary process which is not necessarily originating and diffusing from the global North but actually emerging and developing at the same time across many different geographies and evolving in a variegated way. Gentrification is a complex process of class restructuring of urban space and takes different forms depending on the local context that we are studying. Most literature on gentrification focuses on residential aspects with retail transformations often seen as a secondary, but our study of markets suggests that gentrification is having a significant effect in the retail environment of our cities.

Gentrification is certainly a process affecting the transformation of markets in the cities of Madrid, Leeds, London, Mexico, Santiago and to a lesser extent in our case studies in Quito and Sofia. In Mexico City as well as more generally in Spain, as discussed in Chapter 6, there is a well-established gourmet market model which epitomises this gentrification trend. These spaces are no longer accessible and affordable sources for everyday shopping but have become high-end tourist and leisure destinations.

As highlighted by García et al. in Chapter 7, markets need to be studied in relationship to the neighbourhoods where they are based and the wider urban transformations happening in their cities which results in a nuanced and situated study. For example, in the markets that we studied in Madrid gentrification processes were very different according to the location; not all markets in Madrid are being gentrified – there is what the authors call a 'selective' process. On the one hand, the San Antón Market, based in the gentrified neighbourhood of Chueca was redeveloped (after years of decline and neglect) in 2011 with a new gourmet food offer, restaurants, bars and long opening hours to fit the socio-economic profile of the neighbourhood. On the other hand, not far away, the San Fernando Market is following a very different process which interestingly might eventually lead to a similar outcome. It is based in a traditionally low income neighbourhood rapidly changing into a 'hipster' area. Since 2012, the traders' association, to counter the previous decline of the market, started to offer low rents and diversify the types of stalls in the market attracting young food entrepreneurs and restaurateurs as well business related to the creative economy. Now it is evolving into a hip, alternative gourmet and leisure space. Thus although the markets started from a very different place, they are beginning to converge into a gourmet/gentrified space.

Interestingly, the two different markets map onto to Zukin et al.'s (2009) categories of 'corporate' and 'entrepreneurial' retail capitals which although having different trajectories can have similar effects. The third market discussed in their chapter, Los Mostenses, however seems to be following a very different path for the moment and it was recently described by a newspaper as an '"anti-posh" market: neither gourmet, modern or expensive' (López Iturriaga, 2016).

In La Vega Market in Santiago, Schlack et al. in Chapter 3 have started to see a potential trend towards the gentrification of the market and the process is rather different from the discussed above. The market is owned and managed by the association of traders who are committed to maintaining it as a public space which is open and affordable for all. However, the market has been for some years framed as a tourist and 'foodie' destination by the local and international media and the traders are starting to see more customers demanding high quality and more expensive produce. This could potentially lead the traders to start shifting their offer, neglecting the more traditional function of the market although the trend is not established yet.

Hence we see a variety of gentrification processes affecting markets, sometimes as a result of top-down state-led redevelopments or profit-seeking private developers or even traders and alternative consumption initiatives.

Alternative or hipster markets?

In the introduction of the book we proposed that markets can also become spaces for alternative forms of consumption which can present a challenge to corporate capital and the trends of gentrification discussed above. The Solidarity Market of Bonpland in Buenos Aires presented by Habermehl et al. in Chapter 8 is the most illustrative case. In this small market, customers, producers and traders are directly connected through networks linked to the agro-ecological and the social economy. Mercado Bonpland emerged out of the 2001 economic, social and political crisis in Argentina and is strongly based on the spirit of the social movements that in the wake of the crisis promoted local neighbourhood assemblies and a barter network. At a first glance, we could draw a comparison between Bonpland and San Fernando Market in Madrid, some of whose traders are also introducing fair-trade and locally sourced products and the traders' association is promoting the market as a social meeting space in the neighbourhood with free dance lessons, live music and other cultural activities. Both markets are also located in gentrifying neighbourhoods (Palermo in Buenos Aires and Lavapiés in Madrid). Although we could see similarities between these spaces, their political economy is radically different: San Fernando is analysed by the authors in Chapter 7 as a space in transition, in a trajectory from a period of abandonment to a potential phase of gentrification/hipsterisation. The market seems to be drifting to accommodate demand from new residents and tourists in the neighbourhood. Bonpland however is analysed as an established hub between networks of production which have a long term tradition and have emerged out of social struggles firmly based on the solidarity economy. Bonpland is a relatively de-commodified space and to an extent disconnected to the housing market trends of its neighbourhood while San Fernando cannot escape the market forces shaping the neighbourhood and the city around it.

Thus, although the markets might appear to have some similarities they are embedded in very different local contexts, political economies and struggles. Comparing them is useful in highlighting these differences that a superficial analysis might miss.

Gourmetisation

If a critical analysis can differentiate San Fernando and Bonpland Markets relatively easily, sometimes there can be a fine line between gourmet, organic and farmers' markets and the kind of solidarity and social economies that we have seen in the case of Buenos Aires. The aesthetics and the appearance can be similar and this is something that marketing experts play with when they promote gentrified markets. The consumer is sold the impression that they are taking part in a fairer, more respectful and authentic kind of consumption when in fact their ethics might be no different to a corporate retailer. Unfortunately, alternative forms of production and consumption at scales as different as international fair-trade, farmers' markets, organic and locally sourced food, although they might emerge out of struggles to redress various social injustices, can be incorporated into capitalist logic as a niche option for wealthy consumers (Goodman et al., 2012).

Foodism, the trend for expansion of culinary taste (Johnston and Baumann, 2014), is accelerating the commodification of these potentially alternative forms of production and consumption and markets seem to be a prime place for experimentation with these tendencies. In fact, as we have seen in our case studies, processes of urban regeneration and gentrification are interacting with foodist and gourmet consumption trends. In Kirkgate Market in Leeds, the redevelopment of the market that we discussed in Chapter 9 to upgrade it and attract wealthier customers has included a 'new dining experience, street food cafes serving tasty treats from across the world' (Leeds City Council website, n.d.) and the local authority is now recruiting new traders to sell products such as artisan bread, delicatessen or craft beer. The plans for the major redevelopment of La Merced Markets in Mexico City, discussed in Chapter 2, include a new national gastronomic centre. In Santiago, traders of La Vega market are picking up on a new interest in healthy, fresh and exotic produce by wealthier customers. We also have the self-defined gourmet markets such as San Antón in Madrid or Mercado Roma in Mexico City. As discussed in Chapter 6, the gourmetisation of markets is an international trend where certain markets such as La Boqueria in Barcelona or Borough Market in London have become models for other cities wanting to attract foodie tourists and consumers. In many of our analysed case studies, we have identified a trend for more stalls in markets to be selling gourmet, specialised and delicatessen type products. This 'gourmetisation' trend is reframing markets away from their function as public spaces for the distribution of affordable produce into an elite consumption option.

Zukin (2010) has highlighted how the search for authentic places is driving gentrification in cities. Our research confirms this quest for authenticity in markets which is now becoming intensified by a search for different, authentic and exotic foods. As we have seen already, this is putting pressure on the markets' traditional role of delivering affordable fresh produce.

Contested notions of authenticity and heritage

The search for authenticity in our cities today is triggered by loss of sense of place that globalisation and the standardisation of urban landscape is bringing about. But what and how something is deemed authentic and who has access to it becomes contentious. Indeed, as Zukin has pointed 'whether it's real or not, then, authenticity becomes a tool of power' (Zukin, 2010, p. 3) and so the 'authentic' nature of markets, as we have seen in several of our case studies, has become a source of contention.

The perceived authenticity of markets derives from the fact that they seem to offer us a door into what we imagine might have once been a premodern space before the regularisation, formalisation, globalisation, standardisation and the acceleration of urban life. It is a place where we can reconnect with other humans in a relatively informal space very different from supermarkets, shopping malls and clone shopping streets. It is also a place where we can reconnect with the products we buy, learning about their origin and the process whereby they got to the marketplace. These qualities are appreciated by many users in our case studies but they become exaggerated and marketed when markets are objects of redevelopment and regeneration policies and a target for tourists. For example, in the case of Leeds Kirkgate Market, markets in Madrid or the various gourmet markets mentioned in Chapter 6 in Mexico and Spain, references to these premodern 'authentic' elements are exalted by their promotional websites.

Such commodification of a constructed past has been mainly analysed by heritage tourism studies where the past is 'staged' in search for an authenticity (Halewood and Hannam, 2001). A similar process is taking place in the case of markets. The authenticity and traditional nature of markets is exaggerated and staged as the markets lose their public function as centres of provision for affordable food and basic needs (Pintaudi, 2006) and need to be marketed to new audiences.

Interestingly these pre-modern and traditional qualities can equally be mobilised to stigmatise markets. We have seen this process in action in some of our case studies such as Sofia and Quito, where markets are associated with indigenous populations and culture. In Chapter 10 Eneva discusses how markets have often been discussed in Eastern Europe as 'bazaars', a term that has come to be associated with the oriental marketplace generally 'used to indicate chaotic, disorderly and irrational places of exchange' (Favero, 2010, p. 62). As she explains in the case of the Women's Market in Sofia, a discourse of stigmatisation of this market prior to its redevelopment was accompanied by references to its 'orientalism'; not deemed appropriate for a modern and European Sofia.

Similarly, although in a completely different context, we see how the popular indigenous culture and economy in Quito are associated with the disorderly San Roque market and neighbourhood analysed in Chapter 11 by Garcés and Bedón. What is more, the connection is further made between criminal networks operating in the market and indigenous groups. The San Roque market, which is starting to be reframed as a heritage site due to its vicinity to the UNESCO heritage city centre, is regarded by the authorities as a misuse of such a strategic location. La Merced Market in Mexico City, discussed by Delgadillo in Chapter 2, is subject to a similar discourse. The proposed mega-redevelopment project there is defended by the authorities as a 'rescue operation': rescuing the heritage of the city for uses that are considered more appropriate.

This re-appropriation of heritage in city centres has been discussed by Latin American scholars as a process of 'patrimonialisation', where elite and colonial notions of heritage are imposed upon the built environment erasing in the process any elements that might not be suitable for real estate and/or tourist developments (Hiernaux and Gonzalez Gomez, 2015). As markets lose their original functions as food provisioning spaces and are transformed into leisure and tourist spaces, they start to get patrimonialised, their traditional features commodified as part of an authentic experience not only for tourists but also for occasional shoppers; in these transformed markets other users can be and feel excluded.

Informality

Most scholarship on markets is located within the field of 'informal economy' and focuses on street and informal markets often in the Global South (see for example Bhowmik, 2012; Bostic et al., 2016; Bromley, 2000; Brown, 2006; Brown et al., 2011). Moreover, informality is often spatially identified with slums and street trade. However, in our book we primarily concentrated on formal and regulated markets, mainly indoor and mostly owned and managed by state authorities both in Europe and Latin America and we have found that informality also plays an important role in these kinds of formal spaces. Urban informality of course is a complex issue (Roy, 2005) and this became clear in our study of markets as contested spaces.

Markets often operate in a liminal space between the private and public. They are often regulated by state authorities, but at the same time, market traders run their businesses as private entrepreneurs and establish direct relationships with customers and users. There is a much more informal atmosphere than in a supermarket; rather than standardisation and uniformity we find diversity and unpredictability. Prices are not always fixed or displayed and they vary according to various elements: wholesale price, time of the year or day and the relationship between the customer and the trader.

Informality, as we have seen in this book, is not something happening at the margins or in peripheral spaces of the city; it is at the heart of urban life but at the same time always under negotiation and at risk of being extinguished. This is beautifully articulated in Chapter 11 by Garcés and Bedón's concept of 'urban bustle' which characterises San Roque market. This sense of movement, and to some extent chaos, was often used to describe markets in our case studies as: teeming, noisy, smelly, and in more negative ways as invading, over-spilling or swarming. There is a sense of the informal practices of markets constantly in tension and encroaching on (Bayat, 2013) the more formalised and regulated as well as the real estate market logic of the state and private developers; trying to outgain each other's territory. This is seen in the cases where there are or have been redevelopment projects such as in Leeds, various Madrid and London markets, Women's market in Sofia, La Merced in Mexico or San Roque in Quito. The aim of the redevelopment is to 'reclaim' the informal and undervalued spaces of the markets into the patrimonialised or touristified city; to bring them in line with the land values that have been rising around them.

This tension between informality and formality is negotiated in different ways across the markets that we have seen. In the case of the open air *Tianguis* markets in Mexico City, discussed in Chapter 5, there are very well organised trader organisations that negotiate with state officials about the use of the public space; and in turn trader leaders create codes and rules about the use and exchange of public space within their organisations. These negotiations often crystallise a commitment by the trader organisations to support a particular political party. In other cases the tension is more confrontational: in La Merced, San Roque, London markets and Leeds Kirkgate Market we have seen well organised traders and/or citizen organisations contesting and resisting plans for the regularisation, formalisation and commodification of their markets.

Our book therefore contributes to ongoing debates on the question of urban informalities by showing how in markets across Europe and Latin America informality is very much a central aspect despite these spaces being formally regulated. And as discussed in Roy (2005) our work also shows the important role of state and private capital actors in taming this informality but as we will see below also creating and maintaining it through neglect and abandonment.

Spaces for solidarity and inclusion

The informality that we found in our formal markets is part of the mix of elements that can make them particularly inclusive spaces as we already highlighted in the introduction and as is so evident in our case studies. The marketplace is often regarded by traders, authorities and citizens as an open and accessible public space, much more than a shopping centre. This publicness is discussed in particular by Schlack et al. in the case of La Vega market in Santiago (Chapter 3). The market here is owned by an association of over 2,000 stallholders and it plays an important role in the city as a space open to all, in particular to the most vulnerable. The traders actively support homeless and hungry people offering free food and job opportunities to ex-criminals or marginalised individuals. In particular, it is a space of solidarity and integration for migrants from other Latin American countries. In a completely different space and part of the world, another market is also offering a place for comfort, job opportunities and solidarity for Latin American migrants as discussed in Chapter 4. Seven Sisters or Pueblito Paisa market in North London has become almost a second home for many people; much more than a shopping centre, it's a place where children play after school, an advice centre for new migrants and the centre of a community.

These forms of solidarity and social relations were also evident in the cases of San Roque Market in Quito and Kirkgate Market in Leeds as discussed in Chapters 11 and 9, respectively. In Quito, San Roque Market is a place for the reproduction of indigenous forms of communitarian relationships, expressed in helping each other in the daily trade but also in the maintenance of religious cultural activities. In Leeds, traders of Kirkgate Market practiced an ethic of care towards other traders and customers.

These forms of solidarity and sociability are maintained in the interstices of the formal, regulated and marketised city. Markets have often been out of view or simply tolerated by the state and private capital which has allowed the creation of this particular ecology: affordable rents, softly regulated space, openness, low income and vulnerable users. However, as we have pointed out, this ecology is very fragile and in many cases at risk of disappearing.

Displacement

Markets and market traders, as we have seen in our book, very often live under the threat of displacement. Displacement from central areas in cities of the most vulnerable citizens has now been identified as a central feature of urban development throughout many parts of the world (Janoschka and Sequera, 2016; Lees et al., 2016; Sassen, 2014). In particular, the removal of informal and street traders from public areas and streets in the global South is well documented (Bhowmik, 2012) and has even been recognised by UN Habitat as a problem (Habitat, 2015). However, what we have shown in our book is that regulated and formal markets are now also facing displacement. The process is not as dramatic and visible as the forceful eviction of informal traders from streets but much more of a longer term and gradual process involving various phases and parallel mechanisms. As we saw in the cases studies of Leeds, London, Madrid, Mexico City, Sofia and Quito, the abandonment and/or the marginalisation of the public market is a key aspect that leads to the displacement of traders. The infrastructure is neglected, criminal activities tolerated, the market and the area are stigmatised and a discourse of obsolescence is established in public opinion by the media and authorities. This, to an extent, planned abandonment leads to displacement of markets as the buildings and stalls fall out of repair and traders struggle to keep their businesses. Thus follows the need for intervention, to redevelop, clean or 'recover' the market often led by public authorities but also by private investors.

Redevelopment projects can lead to the displacement of traders and customers in various ways. As we have seen in our case studies, markets often occupy central or valuable land in cities for which state and private actors would like to extract more rent; through a redevelopment project markets get moved somewhere else to make way for other buildings or are kept but rents go up to match higher land values. In the process, traders who cannot afford the new higher rents get left behind. In the San Antón Market of Madrid, during the redevelopment project that took place in 2002, more than half of the traders gave up their stalls to bigger businesses and today there are just twelve market stalls left selling very high quality foods at high prices. In Leeds, the redevelopment project that started in 2014 made it very difficult for dozens of traders to carry on with their businesses due to disruption from construction, relocation costs and higher rents.

In the introduction and through our case studies we have highlighted the direct link between these processes of displacement and gentrification. The aim of many of these redevelopment plans is to change the socio-economic profile of traders and in particular market users to higher-end consumption. In many of our cases, particularly in Sofia, Quito and to some extent in Mexico City and in London and Leeds too, there was a racialised angle to this displacement as markets are/were an area for the concentration of ethnic and indigenous minorities. In the San Roque market in Quito we saw that the authorities quite explicitly wanted to stop and disperse the strong indigenous and popular culture and economy that has been established in the neighbourhood around the market. In all cases there was definitely a class angle: markets were the place of work and shopping for low income citizens.

But in the case of the gentrification of markets, it is not only traders and customers that are displaced but also products and practices. As we have seen in many of our case studies and as discussed above, a trend for gourmetisation is replacing the produce sold in markets from basic daily products and services to much more high-end uses that attract a wealthier clientele.

Contested markets and the right to the city

Finally, an important issue that was highlighted in our cases is that the struggle for markets is often a demand for the right to the city. The right to the city, a phrase originally developed by Henri Lefebvre, has been taken up by urban social movements across the world as a demand for all citizens to have access to basic rights and benefits and for democratic control over resources. For Marcuse, this right to the city is a cry and a demand from the most vulnerable (Marcuse, 2009) and markets, as we have seen in our cases, often serve the marginalised in our cities.

Thus the struggle for markets can become the struggle for a fairer city, a city based on social reproduction, where people have access to affordable food and services, which is not over-regulated and corporatised. In several of our case study markets we have seen traders and market users invoking these rights. In Quito, although not directly mentioned in Chapter 11, a 'Front for the Defence and Modernisation of San Roque Market' led by traders presents the market as a 'home for all' and has developed strategies for the recording and maintenance of popular cultural practices as well as highlighting the role of the market in strengthening the food sovereignty of the city (Frente para la Defensa, n.d.). In London, as was discussed in Chapter 4, many of the market campaigns – particularly those led by market users – see their struggle as a defence of the most vulnerable to 'stay put' against the displacement tide of real estate speculation in the city. In Leeds, the campaign group Friends of Leeds Kirkgate Market, mentioned in Chapter 9, saw the market as a metaphor for the transformation of the whole city centre, towards a more corporate and privatised space.

However, these campaigns are often small and stay disconnected. And despite the evidence we have gathered in this there is not yet a public awareness about the transformation that traditional markets are undergoing. Redeveloped, gentrified and sterilised markets have become tourist attractions all over the world as our search for exotic and authentic experiences and food continues. At the same time, traditional markets – street, covered, informal, itinerant, fixed – selling food and offering a multitude of services still play a central role in cities and provide jobs and affordable produce for millions of people. It is the strain between these contradictory and diverse tensions that is turning markets into contested spaces in our contested cities.

References

Atkinson, R. and Bridge, G. (2004). *Gentrification in a Global Context*. London: Routledge.

Bayat, A. (2013). *Life as Politics: How Ordinary People Change the Middle East*. Stanford, CA: Stanford University Press.

Bhowmik, S. (2012). *Street Vendors in the Global Urban Economy*. London: Routledge.

Bostic, R.W., Kim, A.M. and Valenzuela, A.J. (2016). Contesting the streets: Vending and public space in global cities. *Cityscape*, 18, 3–10.

Bromley, R. (2000). Street vending and public policy: A global review. *International Journal of Sociology and Social Policy*, 20, 1–28.

Brown, A., Lyons, M. and Dankoco, I. (2010). Street traders and the emerging spaces for urban voice and citizenship in African cities. *Urban Studies*, 47(3), 666–683.

Brown, A. (Ed.) (2006). *Contested Space: Street and Livelihoods in Developing Cities*. Rugby: ITDG Publishing.

Favero, P. (2010). Bazaar. In: Hutchinson, R. (Ed.), *Encyclopedia of Urban Studies*. Los Angeles, CA: Sage, pp. 62–65.

Frente de Defensa y Modernizacion del mercado de San Roque (n.d.). Website. Available from: https://frentemercadosanroque.wordpress.com/category/frente-de-de fensa-y-modernizacion-del-mercado-san-roque/

Goodman, D., DuPuis, E.M. and Goodman, M.K. (2012). *Alternative Food Networks: Knowledge, Practice, and Politics*. London: Routledge.

Habitat (2015). *The Informal Sector*. New York: UN Habitat.

Halewood, C. and Hannam, K. (2001). Viking heritage tourism: Authenticity and Commodification. *Annals of Tourism Research*, 28(3), 565–580.

Hiernaux, D. and Gonzalez Gomez, C. (2015). Patrimonio y turismo en centros his-tóricos de ciudades medias. ¿Imaginarios encontrados? *URBS. Revista de Estudios Urbanos y Ciencias Sociales*, 5(2), 111–125.

Janoschka, M. and Sequera, J. (2016). Gentrification in Latin America: Addressing the politics and geographies of displacement. *Urban Geography*, 37(8), 1175–1194.

Johnston, J. and Baumann, S. (2014). *Foodies: Democracy and Distinction in the Gourmet Foodscape*. Abingdon: Routledge.

Leeds City Council (n.d.). Where street food meets market. Leeds: Leeds City Council. Available from: http://www.leeds.gov.uk/leedsmarkets/News/Pages/Where-Street-Food-Meets-Market.aspx

Lees, L., Shin, H.B. and Lopez-Morales, E. (2016). *Planetary Gentrification*. Cambridge: Polity.

López Iturriaga, M. (2016). Los Tesoros de un mercado anti-pijo. *El Pais*, 10 March 2016. Available from: http://elcomidista.elpais.com/elcomidista/2016/03/03/articulo/ 1457020804_692326.html

Marcuse, P. (2009). From critical urban theory to the right to the city. *City*, 13(2–3), 185–197.

Pintaudi, S.M. (2006). Os mercados públicos: metamorfoses de um espaço na história urbana. *Scripta Nova. Revista electrónica de geografía y ciencias sociales*, vol. X, n. 218(81). Available from: http://www.ub.es/geocrit/sn/sn-218-81.htm

Roy, A. (2005). Urban informality: Toward an epistemology of planning. *Journal of the American Planning Association*, 71(2), 147–158.

Sassen, S. (2014). *Expulsions*. Harvard, MA: Harvard University Press.

Zukin, S. (2010). *Naked City. The Death and Life of Authentic Places.* Oxford: Oxford University Press.

Zukin, S., Kasinitz, P. and Chen, X. (2016). *Global Cities, Local Streets. Everyday Diversity from New York to Shanghai.* New York: Routledge.

Zukin, S., Trujillo, V., Frase, P., Jackson, D., Recuber, T. and Walker, A. (2009). New retail capital and neighborhood change: Boutiques and gentrification in New York City. *City & Community,* 8(1), 47–64.

Index